Water Tech

This book unveils how the world in the twenty-first century will need to manage our most fundamental resource: water. It outlines how stakeholders can improve water use in their homes, their businesses and the world.

In particular, it focuses on the role of investors in crafting a twenty-first century paradigm for water. Investors not only drive innovation through direct investment in new technologies, but also by highlighting risk and driving reporting and disclosure within the business community.

Water Tech highlights the business drivers to address water scarcity. These include business disruption, regulatory risk and reputational risk, along with opportunities in the commercialization of innovative technologies, such as desalination and water reuse and treatment. The authors argue that through increased attention on water scarcity (via activities such as reporting and disclosure) we are now accelerating innovation in the water industry. They show how we are just now capturing the true cost and value of water, and this is creating opportunities for investors in the water sector. The text takes the reader through key aspects of emerging innovative technologies, along with case studies and key issues on the path to commercialization. A roadmap of the opportunities in the water sector is presented based on interviews with leading authorities in the water field, including innovators, investors, legal experts, regulatory experts and businesses.

Will Sarni is an internationally recognized thought leader on water stewardship and sustainability strategies, and author of *Corporate Water Strategies* (Earthscan, 2011). Will works with some of the most recognized global brands in developing water stewardship strategies. He is a board member of the Rainforest Alliance, and has worked with several NGOs as an adviser on water-related programs. He is based in Denver, Colorado, USA.

Tamin Pechet is CEO of Banyan Water, a private-equity-funded company using information technology to reduce water costs and risks for large commercial and institutional customers. He is also chairman and co-founder of Imagine H2O, a global non-profit organization spurring water entrepreneurship. He is based in San Francisco, California, USA.

"Water stewardship is in its infancy – yet the possibilities for private industry to drive innovation and support improved performance from the public sector is huge. Private industry has always sought water innovation, but in the past, the pressures were different from those emerging in today's highly branded, globalised and increasingly water stressed world. This book lays out in clear terms why companies need to act and shows how Water Tech will play a crucial role in bridging the internal with external worlds of water management with stewardship practice."

Stuart Orr, Head of Water Stewardship, WWF International

"This book is a welcome, uplifting addition to the water literature. It points out that water problems in fact can be solved, once the risk is properly understood. Thanks to this and other contributions by Will Sarni we are now approaching the point where necessity meets ability, and where water becomes an investment opportunity."

Piet Klop, PGGM

"Water matters to us all, whether CEOs, elected officials or consumers. Two-thirds of the world's population over the next 20 years will experience some type of water shortage. Sarni and Pechet have written an accessible primer on the dynamics of water supply and demand and on the way forward for industry leaders, government regulators, municipal managers and financiers. Much of our profligate water use – including the 70% used by agriculture – is easily reduced through tracking, pricing, efficiency, recycling, reuse and innovation (technology and practice). Industry can lead the way through individual action and precompetitive collaboration with government, financiers, and civil society to address issues of pricing, regulation, and commercialization of new innovations."

Tensie Whelan, President, Rainforest Alliance

"As in energy, addressing the global water challenges of the 21st century will require innovations in technology, investment, and thinking. Sarni and Pechet's book is a highly readable and invaluable guide helping point the way to a new, sustainable water future for the planet."

Clint Wilder, Senior Editor at Clean Edge and co-author,
The Clean Tech Revolution and Clean Tech Nation

"Will Sarni and Tamin Pechet write with passion and optimism about the need to integrate good water stewardship into the heart of business. They show how an increasing number of companies recognizes the need for sustainable water use. The book shows that the path from awareness to actual change will have to go through innovation."

Arjen Y. Hoekstra, professor in Water Management, the Netherlands, and author of
The Water Footprint of Modern Consumer Society

"This book makes a compelling case for why leaders need to better understand our relationship with water, is brilliant in its capture of nuance in water issues around the world, and more importantly it is convincing about the phenomenal commercial opportunity for innovation and technology to contribute to a secure and sustainable future."

Anand Shah, Founder, Sarvajal

"Will Sarni and Tamin Pechet have compiled not only an important book, but also a guide for entrepreneurs, innovators, policymakers and corporate executives. As a longtime water advocate and author on the subject, I can safely say this breaks new ground. Will and Tamin's quest to showcase and discuss the cutting edge tools, practices, and strategies behind the business of water is a wild success. Water is life. And this book identifies new ways we can make it sustainable."

Tom Kostigen, author of The Green Blue Book,
the simple water savings guide to everything in your life

"Sarni and Pechet make it clear that increasingly, water could be a source of failure for companies. There is great opportunity to take an out-of-the-box approach to innovation in water that can lead a more sustainable future for our most critical resource."

Jigar Shah

Water Tech

A guide to investment, innovation, and business opportunities in the water sector

William Sarni and Tamin Pechet

Routledge
Taylor & Francis Group

LONDON AND NEW YORK

earthscan

from Routledge

First published in paperback 2024

First published 2013
by Routledge
4 Park Square, Milton Park, Abingdon, Oxon OX14 4RN

and by Routledge
605 Third Avenue, New York, NY 10158

Routledge is an imprint of the Taylor & Francis Group, an informa business

© 2013, 2024 William Sarni and Tamin Pechet

Publisher's Note
The publisher has gone to great lengths to ensure the quality of this reprint but points out that some imperfections in the original copies may be apparent.

British Library Cataloguing-in-Publication Data
A catalogue record for this book is available from the British Library

Library of Congress Cataloging-in-Publication Data
Library of Congress Cataloging-in-Publication Data
Sarni, William.
Water tech : a guide to investment, innovation and business opportunities in the water sector / William Sarni and Tamin Pechet.
pages cm
Includes bibliographical references and index.
1. Water resources development—Technological innovations.
2. Water-supply—Economic aspects. 3. Risk management. I. Pechet, Tamin.
II. Title.
HD1691.S28 2013
333.91—dc23
2013010778

ISBN: 978-1-84971-473-0 (hbk)
ISBN: 978-1-03-292668-1 (pbk)
ISBN: 978-0-203-12729-2 (ebk)

DOI: 10.4324/9780203127292

Typeset in Goudy
by FiSH Books Ltd, Enfield

William Sarni:
This book is dedicated to my wife Maureen, and my sons James, Thomas, and Charles. They inspire me to contribute what I can towards creating a better world.

Tamin Pechet:
For my father, with gratitude.

Contents

List of illustrations

Figures

Boxes

Foreword

Living as I do in London, England, I take it for granted that if I turn on a tap in my home – or anywhere else in the city – I will be rewarded with as much water as I want and that it will be clean enough to drink. For over a century Londoners like me have relied on largely Victorian-era water tech to supply our water and wash away our waste, and it has been easy to imagine that the city's water challenges have been permanently solved.

That belief is now being challenged. Rising demand for water from a growing population and an apparent increase in the variability of rainfall is stretching the city's aging infrastructure to its limits. After two successive dry winters we are in the midst of a drought, and although it may prove only mildly inconvenient – restrictions are currently limited to a hosepipe ban for gardeners – it does presage a more challenging future.

London's experience is by no means unique. Massive upgrades in infrastructure are needed across much of the developed world, with the American Water Works Association estimating that the cost of upgrading and expanding US drinking water infrastructure alone will be $1 trillion over the next 25 years. Even more sobering is the fact that 780 million people in developing countries still lack access to safe drinking water, while 2.5 billion people lack access to improved sanitation facilities.

The challenge for the water industry does not end there. Global demand for food and energy is projected to increase by 50 percent by 2030, and both require vast quantities of water. Businesses too are thirsty, and rely on the provision of water in the right quantity and quality at the right time, which frequently puts them in competition with communities and ecosystems for what is a finite, unsubstitutable, life-sustaining resource. This increasing competition for water poses a very real risk to companies. Although this risk is not yet widely acknowledged or understood, investors are beginning to take note as companies post reduced earnings or losses resulting from disruptions to their operations or supply chains caused by drought or flood, from fines and litigation relating to pollution incidents, from tighter regulations that restrict access to or increase the price of water, or from reputational damage that reduces demand for their products.

It was to raise awareness and understanding of these business risks (and related opportunities) and to encourage action to manage them through better corporate water stewardship that the Carbon Disclosure Project (or CDP) launched its water program in 2009. This year, on behalf of 470 banks, pension funds, asset managers, insurance companies, and foundations, together representing $50 trillion in assets, CDP is seeking disclosure from almost 650 of the world's largest companies on their water usage, their water management and governance, and the risks and opportunities that water presents to them.

The model is powerful. CDP's experience with energy and carbon has shown that companies manage what they measure. Once armed with information about their energy use and carbon emissions, and an understanding of the risks and opportunities that these present, the logical next step for companies is to develop strategies to reduce their emissions, manage their risks and seize their opportunities. These strategies in turn have been an important spur to clean tech innovation as companies seek more efficient and sustainable processes and business models. In 2011 alone companies reported almost 10,000 emissions reduction activities through CDP, spending billions of dollars on solutions, many of which are expected to pay for themselves in under three years.

We are already seeing a similar pattern emerge with water. Many companies have begun to take that first step of measuring and reporting their water usage, and there are encouraging signs that leaders are developing a sophisticated understanding of the value of water to their businesses and implementing strategies for a water-constrained world. A steady trickle of water tech solutions is beginning to emerge, from desalination and nanotube filters to bolster supplies, to drip irrigation, recycling, and smart meters to manage and monitor demand, to new techniques to recover high-value resources from waste streams and to clean polluted discharges.

I fully expect this trickle to become a torrent. Water is the lifeblood of ecosystems, communities, and the global economy, and while the challenges in keeping it flowing are huge, so too are the opportunities. A new era of water tech beckons.

Marcus Norton
Head of Water, CDP
May 2012

Foreword

A news report said, "they may have found water on the moon."

Water on the moon? Why is that newsworthy? Because water means there is the possibility of life.

Here on Earth, we are increasingly corrupting that possibility. Water we take for granted. Water we misuse. Water we mismanage. Water we do not value.

All of humankind needs water to survive. And as individuals we can do our share to preserve and conserve this critical natural resource. But that isn't enough.

It is the business of water where innovations, technologies, and facilities can be devised to promulgate water security. The problem with water is that we can't make more of it. Water consists of molecules – two hydrogen and one oxygen – that incredibly come together and break apart, then come together again, over and over throughout the Earth's biosphere.

Almost to the drop the exact same amount of water has existed on this planet since the time of dinosaurs. We the people, however, have mushroomed in terms of population, and spread throughout the world. This directly affects the world's water supply in two ways: (1) there are more people reliant on freshwater, and (2) water must be transported to our more disparate population. Indirectly, the ramifications of our use harm supplies too: pollution can infect water sources.

From a business perspective, the problems and opportunities loom large. That's why this book is so important.

We live in the age of technology, where lasers can be used to purify water, increasing available supplies; where infrared mapping systems can identify previously hidden sources closer to more densely populated areas; where whiz-bang filtration systems can make toilet water clean again. Meanwhile, sensors can allocate water use more effectively (i.e. rain sensors), and water meters can educate and better inform water consumers.

Therefore, water technology is exigent to our survival: Fully two-thirds of the world's population over the next 20 years will experience some type of water shortage. A child under the age of five dies every five seconds from a water-related illness. There is not enough water to meet our growing food and fuel needs.

So what are we to do? Innovate. That is what we humans do and have done over the course of history. We figured ways to capture and store water, freeing us from the shores of lakes, streams, rivers, and reservoirs. We figured ways to irrigate, to free us from the inconsistencies of rain. We even figured ways to transport water into our homes and send it away via sewers. Now it's time to innovate once more. It is not enough to adapt to existing conditions, mainly because we cannot; that is a losing proposition for humankind. We won't survive.

No; "that challenge is one that we are willing to accept, one we are unwilling to postpone, and one which we intend to win." President John K. Kennedy said those words in his famous speech just before man landed on the moon for the first time. And it is with that reverence and resolve that we must explore the domain of water. Indeed, it is with another quote from Kennedy that this book begins.

The mission of what is set forth in these pages is water technology in the twenty-first century. The following pages are filled with facts, figures, stories, insights, data, and information for any individual, any business, or any government seeking to comprehend the way to a better water future.

Tom Kostigen
Author of *The Green Blue Book: The simple water savings guide to everything in your life*
June 2012

Preface

My *Corporate Water Strategies* (Earthscan, 2011) laid out the landscape of how water scarcity represents a business risk and how companies are addressing these risks through water stewardship strategies and developing an understanding of the true value of water. Much progress has been made in addressing water risk in the two years since that book was published.

Progress has come in: new water footprinting and risk mapping; reporting though the CDP Water Program (formerly CDP Water Disclosure); guidelines and tools on collective action; new partnerships between non-governmental organizations (NGOs), companies, and governments; and technology innovation. Most importantly, the private and public sectors are having a real discussion on the value of water. Water is essential for human life, ecosystems and economic activity and as such must be valued accordingly. This rethinking of the value of water is shaping public policy and business decisions, and driving water tech innovation.

Water Tech focuses on progress in technology innovation and builds on some of the ideas and innovations discussed in *Corporate Water Strategies* – low energy water treatment, water reuse and recycling, and distributed water treatment, to name a few. *Water Tech* is the next chapter in telling the story of how we are meeting the global need for water. It chronicles how companies, countries, entrepreneurs and investors are addressing water scarcity through the development of innovative technologies.

This book also brings the unique perspective of my friend, Tamin Pechet, who has been tireless in shaping the world of water tech through the creation of Imagine H2O and Banyan Water.

Technology innovation is part of the solution in providing clean water and sanitation to an ever increasing global population, ensuring there is water for energy and food to support this global population, economic growth, needs of ecosystems and the cultural and social requirements of humanity. Water tech coupled with changes in public policy, new business models, incentives, and collaboration can meet the diverse and ever-increasing need for freshwater.

I do not believe in "business as usual," and remain hopeful that by shining a light on water stewardship strategies and technology innovation we will not experience the projected 40 percent shortfall of water and 47 percent of the population to experience water scarcity by 2030.[1] We are in a position to shape the future – abandon the notion of business as usual, and embrace innovation in new technologies, policies, and thinking regarding water.

Water is our *shared finite resource*. It is up to the public and private sector to value this resource and ensure we all have adequate water to support our ever-increasing needs.

We are all on this journey towards achieving a common goal – access to clean water and sanitation, water for economic and ecological needs.

I hope this book inspires you to join in achieving this goal.

William Sarni

Note

1   *Charting Our Water Future, Economic Frameworks to Inform Decision-Making*, 2030 Water Resources Group Report, 2009.

Preface

During summers, one of my childhood chores was pouring Clorox into the water tank at my mother's home in Bermuda. The island's pastel-painted houses use ridged white limestone roofs to neatly direct rainwater, the only available fresh-water source, into on-site storage tanks. Our tank was adjoined to the kitchen. It was easy, and unnerving, to imagine the blue bleach moving the few feet from the tank to tap. My father, a chemist, supervised the water treatment. At some point he would signal "enough," after a thumb-lick calculation of bleach concen-tration. He gave us fluoride pills to keep our teeth strong, since there was no municipal supply to dose our water. And we kept showers a little shorter than we wanted. In hindsight, we operated our own little water plant.

My family knew the water in Bermuda made me sick, but there was no way around it. Yet, unlike hundreds of millions of people who faced water scarcity and contamination, my Bermuda experience was temporary, and an accepted down-side of an otherwise perfect vacation. If I got a little sick, or kept showers short, it was a choice. Summer would end, and I'd return to Boston, with limitless water provided by a responsible water and sewer utility.

Years later, while working at Goldman Sachs & Co., I learned of the business opportunity in water and thought back to my childhood experiences in Bermuda. I remember asking, "what's a water business?" I thought you either dealt with water on your own, as we did in Bermuda, or a town, like Boston, supplied it nearly for free. I was shocked to learn that water industry revenues reached the hundreds of billions.

Suddenly, I saw an opportunity to seek profit doing something that had personal meaning to me. But if I didn't even know that water was a business, did other industry outsiders? And if I had so little awareness, who was solving the types of problems I experienced in Bermuda?

A decade later, water remains an incredible opportunity to find meaning and money. A pervasive lack of awareness of the opportunity to profit from solving water problems plagues all water stakeholders. I have witnessed a recent crescendo of interest in water among businesses, investors, and consumers. Businesses, cities, farms, and homeowners have begun to recognize that the way we manage water today cannot be the way we manage water tomorrow. And yet, despite signs of real change coming for one of the world's largest industries, the biggest competitor for most water innovation remains inertia.

I wanted to write this book for readers who might not know how important water is, and even more for readers who know water's importance and are ready to act. Each of us is now deeply affected by water issues, some of which are clear and some harder to see. And each of us can affect those water issues, not just by innovating ourselves, but also as customers, company influencers, voters, home-owners, and citizens.

I had the honor of writing the preface to William Sarni's *Corporate Water Strategies* (Earthscan, 2011), and Will and I had so much fun with it that we decided to write this together. We hope that this book becomes a living document, with reader interaction online at the book's website (www.watertechbook.com), and that it inspires you to take advantage of the water opportunity.

Tamin Pechet

Acknowledgements

I am hooked on writing. It doesn't necessarily come easy at times, but having a voice on issues such as sustainability and water stewardship is an increasingly important part of my life. For me it is part of "living in the solution," as my friend Deanna Turner always says.

Writing for me would be a nearly impossible endeavor without the help and support of friends, family, and colleagues who provided constant encouragement and support as the manuscript progressed. As always, I will never be able to find the words to adequately thank my wife, Maureen Meegan, who provided endless support and encouragement to take on the project and keep writing. She sacrificed precious weekends while I worked on the manuscript, and I could not have written this book without her. My sons, James, Thomas, and Charles, continue to provide encouragement for me to write, and are now asking about the next book. They have matured into exceptional men, and are also voices evangelizing the value of sustainability.

Thanks to my sister, Celeste, who is one of my most vocal supporters, and to my parents, Josie and Mike, who instilled in me a love and curiosity for life, a strong work ethic, and the belief that anything is possible. As always, thanks to my Aberman, Casey, Domijan and Zelkovich extended families, and my nieces and nephews, who provide ongoing encouragement.

And thanks to Hillary Mizia, Tom Kostigen, and Deanna "Drai" Turner for making significant contributions in helping me with the research, drafting and editing text, and preparing the graphics. Most importantly, they provided invaluable advice and perspective when it was critically needed.

My thinking about water stewardship strategies and water tech benefited enormously from my conversations with those who are working on addressing the global challenge of water scarcity on a day-to-day basis. Everyone was generous with their time and support, and provided valuable insight on the emergence of water tech innovation.

We would both like to thank for their contributions to the book: Marcus Norton, CDP; Sheeraz Haji, Cleantech Group; Dan Bena, PepsiCo; Stuart Orr, WWF International; Emily Ashworth, a global information technology executive; Doug Henston, a cleantech entrepreneur and former CEO of Solix; John

Dickerson, Summit Global; Tom Pokorsky, Aquarius Technologies; John Schroeder, Marmon Water; Augie Rakow, Orrick; Lang McHardy, Vested IP; and Rebeca Hwang.

And a very special thanks to Tim Hardwick from Earthscan, who once again provided me with the opportunity to write this book, and offered guidance, encouragement, and an enormous amount of patience along the way, and to my co-author, Tamin Pechet, who provided invaluable insight and perspective on water tech and how to succeed as an entrepreneur. I learned much from him.

William Sarni

Thank you to my co-author, Will, for inviting me to write with him, and for his calm and pragmatic approach to writing and to his work in water.

I owe the opportunity to write this book in large part to the Banyan Water and Imagine H2O teams. Each person involved with those organizations – employees, investors, customers, board members, partners, and helping friends – gave of themselves, often at great risk, for the opportunity to change the world of water. Our work together has taught me most of the water knowledge I have to share in this book.

Thanks also to the many water industry leaders who helped me when I first sought a way into the business. Their passion for their work, interest in fresh approaches, and willingness to share and help one another bodes well for our water future.

Thank you to my family. My parents, siblings, nephews, and in-laws have made life easy and meaningful. Most of all, thank you to my wife, Nikki, for making writing this book, and everything else we do, feel important, inspiring, and fun.

Tamin Pechet

Author biographies

Will Sarni is an internationally recognized thought leader on water stewardship and sustainability strategies based in Denver, Colorado, and a frequent speaker for corporations, conferences and universities. He is the author of *Greening Brownfields: Remediation Through Sustainable Development* (McGraw-Hill) and *Corporate Water Strategies* (Earthscan). Will is a board member of the Rainforest Alliance and has worked with several NGOs as an adviser on water-related programs.

Will has worked for some of the most recognized global brands on developing and implementing corporate-wide sustainability strategies and broad-based water stewardship programs. He has a creative approach in developing and implementing high-value sustainability programs and integrating diverse business and technical issues related to resource management.

Tamin Pechet is CEO of Banyan Water, a private-equity-funded company using information technology to reduce water costs and risks for large commercial and institutional customers. He is also chairman and co-founder of Imagine H2O, a global non-profit spurring water entrepreneurship through innovation prizes and a water business accelerator program. He is a member of the board of directors of Lux Research, a leading provider of research and analytics on water and other science-based innovation markets.

Tamin previously worked as a venture capitalist at Catamount Ventures, where he invested in technology and sustainability companies, and as an investor at Goldman Sachs, where he helped launch a new energy subsidiary.

He is a frequent speaker on water business issues. Tamin holds an AB from Harvard University and an MBA from Harvard Business School, where he was featured in two recent case studies taught on water.

Part I

Innovation

Anyone who can solve the problems of water will be worthy of two Nobel prizes – one for peace and one for science.

(John F. Kennedy)

There is an air of frustration in some of my (Sarni's) conversations in Marseille, France, at the 2012 World Water Forum. Figuring out how we accelerate collaboration on water conservation projects within the watersheds in which we operate is part of these conversations. Not a heated debate; instead a genuine desire to deploy resources quickly to collaborate on a wide range of water projects – water efficiency, water conservation, infrastructure, and capacity building, to name a few. The consensus: collectively, we need to move fast.

These are not global water non-governmental organizations (NGOs) discussing watershed conservation projects. Instead these are leaders from multinational companies representing their CEOs in addressing the global and local challenges of water scarcity and water quality. Not just a concern about how water related issues could impact their businesses, but how these issues impact a wide range of stakeholders – civil society, consumers, customers, employees, and other businesses.

Why would CEOs care about collaboration on water projects to address these issues? CEOs (and, as a result, their chief sustainability officers) care about water more than you might think.

The answer comes from Peter Schulte and Jason Morrison from the Pacific Institute:[1]

They care because water scarcity means that there may not be enough water to produce their goods. Water pollution can lead to great costs to treat water to a level suitable for production or possibly strict regulations. A lack of access to clean water and sanitation for communities may mean that company water allocations are curbed in favor of these more pressing needs or that the company is perceived as being complicit in this lack of access. Ineffective public water management may mean that water is not delivered to a company consistently or reliably. Water is a shared resource and we need to find ways to share it equitably or we all suffer.

One of the many stakeholder meetings was a two-day meeting of the CEO Water Mandate.[2] Launched in 2007 by the UN Secretary-General, the CEO Water Mandate is an initiative of the UN Global Compact[3] – operated in collaboration with the Pacific Institute[4] – designed to assist companies in the development, implementation, and disclosure of water sustainability policies and practices. The Mandate produces research that identifies and provides guidance on water-related business challenges and convenes multi-stakeholder working conferences whereby companies and their stakeholders discuss what it means for a company to be a responsible water steward. As of 2012, the Mandate is endorsed by more than 80 companies from a wide range of industry sectors and geographies.

Typically, corporate water management improvements, if present at all, have focused on water use efficiency and reducing water pollution caused by the company. The Mandate and its endorsers are committed not only to these crucial internal improvements, but also to developing and implementing new pathways with which companies can encourage and contribute to the sustainable water management of their supply chains and the watersheds in which they operate.

This expanded view of corporate water stewardship focuses largely on new ideas about how companies can relate to and partners with others to support sustainable water management, namely the concepts of policy engagement and collective action. These emerging strategies are based on two fundamental realities that shape water-related business risks:

- Often the greatest water-related business risks stem from conditions outside of the company fence line, such as water scarcity, poor ambient water quality, insufficient water resources management, inadequate infrastructure, climate change, and others, over which companies have limited influence.
- The same water-related conditions that create risk for business also create risk for communities, the environment, and governments alike. This shared risk creates an incentive for shared, collective response.

Business engagement with water policy, if implemented effectively and responsibly, allows companies to mitigate water-related business risks by encouraging more sustainable water management (especially by means of supporting and enriching government's management capacity) outside their fence lines. Collective action enables companies to partner with others in order to combine resources (e.g. funding, expertise, local knowledge, and innovative practices) to promote shared water-related goals. Taken together, these strategies enable companies to think more comprehensively about the root causes of and most effective solutions for society's critical water challenges.

The CEO Water Mandate meeting in Marseille, France started off with an overview of two key projects, one of which we will now highlight – an update on the development of a Water Action Hub. The "Hub" is an online tool designed to facilitate collaboration between stakeholders interested in addressing a wide range of water issues that are important not only to CEO Water Mandate members, but to a wide range of stakeholders – the public sector, civil society, NGOs, consumers, customers, and supply chain partners for these companies.

The Hub will fill an important gap in how stakeholders are attempting to connect to each other to work on water efficiency, water conservation, and public policy projects, to name a few. The Hub will allow stakeholders to go online and enter information necessary to facilitate a connection – topics of interest, geographic areas of interest, along with key profile information. The conversation in Marseille addressed the "nuts and bolts" of how to build the platform and how to facilitate constructive engagement with interested stakeholders. Much progress was made, building on the enthusiasm and commitment of the project sponsors, technical advisors, advisory committee and other interested parties.

Why bring up the CEO Water Mandate and the Hub? The CEO Water Mandate and projects like the Hub are designed to tackle the complex challenges of managing the ever increasing demands on water supplies – access to clean water, sanitation, water for agriculture and industrial uses, and ecosystem needs. Failure to address access to clean water and sanitation results in disease and deaths, negative impacts to economic growth and failure to meet the needs of agriculture and energy.

The Hub represents innovation in new partnerships, collaboration platforms and processes (Water Futures Partnership[5] is an example of such partnerships) along with changes in public policy, water governance and reporting/disclosure initiatives (such as the CDP Water Program).

It is clear that *innovation is not just about technology.*

Innovation in technology, public policy (including pricing) and partnerships will be needed to solve the ever-increasing challenges of water scarcity and water quality. Innovation in the water industry is coming from all angles and a wide range of stakeholders.

If you picked up this book you likely care about how innovation can address the challenges in providing access to clean water and sanitation together with sustainable water for industry and agriculture. Unfortunately, the statistics and projections for water are currently troubling:

- "It is estimated that two out of every three people will live in water-stressed areas by the year 2025. In Africa alone, it is estimated that 25 countries will be experiencing water stress (below $1,700m^3$ per capita per year) by 2025. Today, 450 million people in 29 countries suffer from water shortages."[6]
- "Clean water supplies and sanitation remain major problems in many parts of the world, with 20 percent of the global population lacking access to safe

drinking water. Around 1.1 billion people globally do not have access to improved water supply sources, while 2.4 billion people do not have access to any type of improved sanitation facility."[7]

- "About 2 million people die every year due to water-borne diseases from fecal pollution of surface waters; most of them are children less than five years of age. A wide variety of human activities also affect the coastal and marine environment. Population pressures, increasing demand for space and resources, and poor economic performance can all undermine the sustainable use of our oceans and coastal areas."[8]
- More than 3.4 million people globally die each year from water-related disease, of which 1.5 million are children under the age of 5.[9]
- Major regions of the world are facing severe drought conditions (when this book was being written about 68 percent of the US was considered under severe drought conditions.[10]

As illustrated in Figure P1.1, water scarcity is projected to result in major areas of the world experiencing water stress or scarcity by 2025.

The increased demand for freshwater is expected to result in an approximately 40 percent shortfall of water by 2030 (Figures P1.2 and P1.3). The impacts to the global population can be gauged by comparing the UN Millennium Development Goals versus how much progress is made against these goals (Figures P1.4 and P1.5).

While we will discuss this in more depth later in the book, it is worth mentioning that water scarcity and droughts are related but different issues. Droughts are weather-related (short-term trends) or climate-related (long-term trends), and water scarcity can occur in the absence of droughts. For example, when this book was being written (2012) the US was in the midst of the worst drought in decades, impacting agricultural and energy production. In contrast, Singapore, which has no shortage of rainfall, is impacted by water scarcity. Singapore is about 680 square kilometers in area, with a population of about 4 million, and is essentially an urbanized country with few natural resources. As a result, Singapore is facing a serious shortage of water resources with demand at about 1.4 million cubic meters daily, but domestic resources only meet about 50 percent of that.[11] Singapore has responded to this long-term water scarcity challenge by developing a strategy for sustainable water management that includes investment in water tech.[12]

We are inspired by the success stories of places like Singapore, Israel, Australia, and South Africa. This book was written because we believe we can address the need for sustainable water management through technology and partnership (referred to as "collective action") innovation in the water industry.

We spend our days on a range of issues, from helping companies develop water stewardship strategies (Sarni) to launching a water technology company (Pechet) and co-founding Imagine H2O (Pechet). Our perspective and inspiration is shaped by interactions with investors, entrepreneurs, and some of the

Water availability in 2000

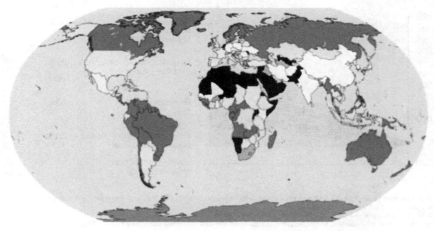

Water availability in 2025

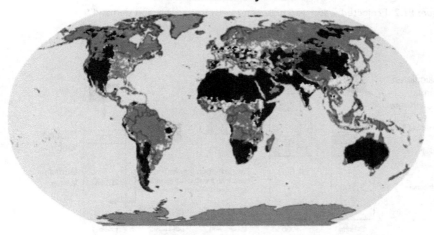

■ **Extreme scarcity <500** (m³/person/year)

Figure P1.1 Water supply per river basin (2000 and 2025)[13]

world's leading companies in the food and beverage, apparel, energy, and oil and gas sectors. As a result, the perspectives in this book will decidedly be a private sector view of the challenges and solutions. This is the world the authors operate in and there are exciting technology developments from the private sector.

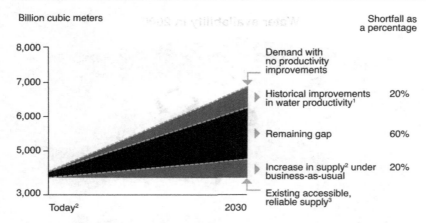

1 Based on historical agricultural yield growth rates from 1990-2004 from FAOSTAT, agricultural and industrial efficiency improvements from IFPRI
2 Total increased capture of raw water through infrastructure buildout, excluding unsustainable extraction
3 Supply shown at 90% reliability and includes infrastructure investments scheduled and funded through 2010. Current 90%-reliable supply does not meet average demand

Adapted From: Exhibit II, 2030 Water Resources Group (2009) *Charting Our Water Future: Economic Frameworks to Inform Decision-Making*

Figure P1.2 Projected water gap between raw water supply and demand[14]

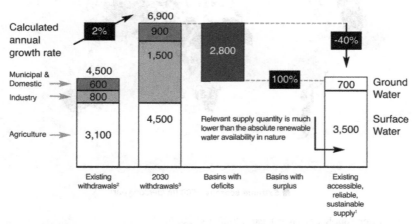

1 Existing supply which can be provided at 90% reliability, based on historical hydrology and infrastructure investments scheduled through 2010; net of environmental requirements.
2 Based on 2010 agricultural production analyses from International Food Policy Research Institute (IFPRI); considers no water productivity gains between 2005-2030.
3 Based on GDP, population projections and agricultural production projections from IFPRI impact-water base case.

Adapted From: Exhibit 1, 2030 Water Resources Group (2009) *Charting Our Water Future: Economic Frameworks to Inform Decision-Making*

Figure P1.3 Global gap between existing accessible reliable supply and 2030 water withdrawals, assuming no efficiency gains[15]

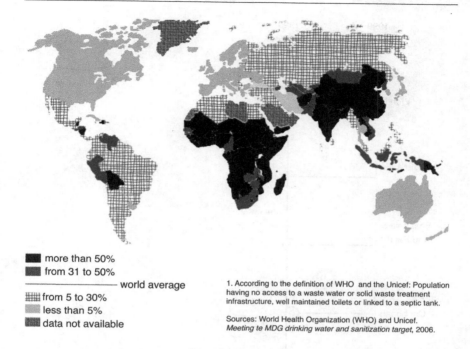

more than 50%

from 31 to 50%

———————————— world average

from 5 to 30%

less than 5%

data not available

1. According to the definition of WHO and the Unicef: Population having no access to a waste water or solid waste treatment infrastructure, well maintained toilets or linked to a septic tank.

Sources: World Health Organization (WHO) and Unicef. *Meeting te MDG drinking water and sanitization target*, 2006.

Figure P1.4 Meeting the Millennium Development Goals drinking water and sanitation targets[16]

As my (Sarni) friend, Deanna Turner (the designer for the cover and graphics in this book), put so well, "we are living in the solution."

We do not believe in "business as usual."

Ensuring business as usual does not materialize will require:

- accelerating adoption of innovation in the water industry;
- supporting public and private sector collaboration such as the CEO Water Mandate Water Action Hub;
- inspiring the next generation of water tech leaders; and
- enabling the social and political transformation required to support water tech innovation.

This book is intended to be a living document, as it would be impossible to keep pace with the rapid developments in water technology and public policy. Instead we will use this book to frame the issues and provide a sense for how people are addressing water challenges through technology innovation. We will connect with the ecosystem of stakeholders tackling innovation in the water industry though our website (www.thewatertechbook.com).

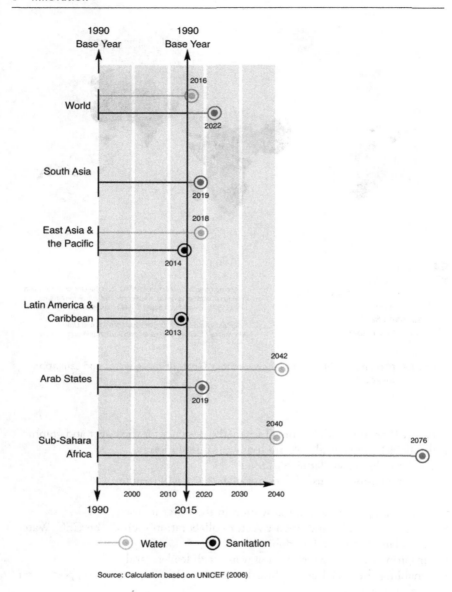

Figure P1.5 Summary of progress towards Millennium Development Goals by region[17]

The book is structured to provide overviews of the key challenges and insights from stakeholders engaged in solving complex water problems – a diverse group of stakeholders – businesses, venture capitalists, entrepreneurs and NGO leaders.

- Part I will provide context for the water technology challenge – the drivers for innovation and what innovation looks like.
- Part II will provide an overview of the opportunities in water tech.
- Part III will provide examples of the landscape as to how technologies go from an idea to commercialization by highlighting initiatives by innovators, investors, legal, regulatory and marketing leaders.

This book will show you what innovation looks like and how there is a new paradigm for water: *"21st Century Water."* This 21st Century Water paradigm is about moving beyond traditional solutions. It will explore how prize competitions such as the Hult Challenge,[18] Imagine H2O,[19] X-Prize,[20] and the Shell Eco-Marathon[21] foster innovation, and finally, how to navigate the challenges of bringing great ideas to the market to solve economic, environmental and social problems. It will provide perspectives from those at the center of this innovation.

We will present the opportunity for World Water Day to be every day, not just a once a year event to raise awareness of the water challenges we face.

Notes

1   Personal correspondence for this book.
2   http://ceowatermandate.org.
3   www.unglobalcompact.org.
4   www.pacinst.org.
5   www.wwf.org.uk.
6   www.unep.org/dewa/vitalwater/article186.html.
7   Ibid.
8   Ibid.
9   www.water.org.
10   http://droughtmonitor.unl.edu.
11   Baumgarten (1998).
12   The World Bank (2006) *Dealing with Water Scarcity in Singapore: Institutions, Strategies, and Enforcement.*
13   www.unep.org/dewa/vitalwater/article75.html.
14   Adapted from 2030 Water Resources Group (2009) *Charting Our Water Future: Economic Frameworks to Inform Decision-Making*, Exhibit II
15   Ibid.
16   Adapted from World Health Organization and Unicef (2006) *Meeting the MDG Drinking Water and Sanitization Target.*
17   Adapted from Unicef (2006).
18   www.hult.edu.
19   www.imagineh2o.org.
20   www.xprize.org.
21   www.shell.com/home/content/ecomarathon.

Chapter 1

The value of water

Imagine you are the CEO, CFO, chief sustainability officer (CSO) or on the board of a company. You don't know much about water or if it is important to your business.

If you are in one of these positions, you should be asking the following questions.

- Is water a critical resource for my company as part of my direct operations, supply chain and/or through product use?
- If yes, is water scarcity a risk to my company (physical, regulatory and reputational)?
- Does water represent a business opportunity through new products or services?
- How much does water cost my company and what are the pricing trends?
- More importantly, what is the *value* of water to my company?
- Does my company have a water strategy?
- What is my company's performance in water efficiency, reuse, recycling, stakeholder engagement and reporting?
- Do I know the stakeholders that would care how I use water?
- Do these stakeholders know how we are managing water?
- How are we communicating our water stewardship strategy to investors, NGOs and other stakeholders?
- How does my water use relate to my energy use and greenhouse gas emissions?
- Does our water use currently or potentially impact energy and agricultural production within the watersheds in which we operate?

So, let's say you have answered yes to the first two questions. You may still not know how much and how it is used throughout your supply chain, direct operations and through product use. You may believe that since the price of water and discharge costs are so low, why should I care? Your facility managers are telling you that the company is a very efficient user of water compared to your competitors, so why should we worry? Or, your environmental, health, and safety manager (EH&S) says you are in compliance with all current regulations and there are no regulatory related risks.

As the CEO, CFO, CSO or board member (to name a few), you should be concerned.

In particular, the CFO should worry. The role of the CFO in managing sustainability issues is increasing, according to a recent GreenBiz and Ernst & Young survey.[1] Historically, the role of the CFO in sustainability has not taken center stage. Instead the CSO or EH&S manager has led sustainability initiatives, but this is changing. CFOs are getting involved in the management, measurement and reporting of the companies' sustainability activities and increasingly water is one of these key issues.

According to this survey, about one in six respondents (13 percent) said their CFO was "very involved" with sustainability, while 52 percent said the CFO was "somewhat" involved. This means that about 65 percent of CFOs are now engaged in sustainability issues. Respondents cited cost reductions (74 percent) and managing risks (61 percent) as two of the three key drivers of their company's sustainability agenda along with increased monitoring of shareholder resolutions.

The GreenBiz report references the impact of the CDP Water Program report on increasing awareness of water related risks and opportunities (see Marcus Norton's Foreword to this book). The 2012 CDP Water Program Global 500 report found that most respondents understood the business risks and opportunities from water related issues (more details on this later in the book). The opportunities reported by the respondents range from the savings realized by using less water to potential new products and services.

What is the value of water to businesses and how do stakeholders view water risks and opportunities? Let's take a look.

The stakeholders

Who cares about innovation in water – a "360 degree view"

Everyone cares about water and relates to water; we can't do without it and there is no substitute.

- *Investors* – For investors, whether private equity, socially responsible investment (SRI) funds, or banks, the questions become: (1) are my investments at risk from water scarcity (or resource scarcity in general)? And (2) are there investments in companies to address water scarcity and access to clean water and sanitation? Investors increasingly care, as reflected in the increase in the number of reports from financial institutions and investor groups on water risk and water tech opportunities. Expect this to continue as savvy investors develop a deeper understanding of potential water scarcity risks to their investments (several publicly traded companies have already identified water as a material risk to their businesses) and recognize that there are opportunities for investment in water tech.

- *Business* – Water-intensive industries such as food, beverage, apparel and power generation understand that water scarcity has or can impact their business through business disruption and reputational risk. We are now seeing companies that might not be considered water-intensive, such as the manufacturing or retail sectors, mapping water risk and developing mitigation programs. It is a case of increasing awareness of water risk and opportunities with global companies and value chains (upstream supply chain and downstream product use). It is also worth noting that consumer product companies are now developing an understanding of how much water it takes for consumers to use their products – from toothpaste to shampoo to household cleaning products. No water, no product use (or less product use).
- *Domestic use* – Every resident understands what happens when disaster strikes and you are without water and power. Run to the store (if you can) for bottled water. During periods of drought domestic users of water can feel the impact of water restrictions – no watering your lawn, washing the car, and so on. Domestic users are also beginning to feel the slow upward trend in water prices (arguably not much) which may draw attention to how wasteful we are in water use – in the US we flush our toilets with potable water and have the greatest per capita water use in the world. This is hardly valuing water. There is progress in the domestic sector in the adoption of green building practices where water (and energy) efficiency is built into residential design. However, more has to happen with regards to the efficient domestic use of water (point of use treatment, water efficient fixtures, etc.).
- *Agriculture* – On the average, the agricultural sector is the largest user of water globally with roughly about 70 percent of global water use.[2] Within this sector there is a growing awareness of the need to improve water efficiency through technology applications (drip irrigation where feasible), drought resistant crops and salt (or brackish water) tolerant crops.
- *Water utilities* – How do water utilities operate with the economic reality that there are few if any funds for expansion of infrastructure, a push back on price increases and increasing operating costs? The water utility sector is deeply challenged in meeting the water needs of the domestic and industrial sectors. These utilities will need to be paid a fair price for delivered water services and will need to be able to adopt technologies that promote efficient use of water resources (smart water meters for example). Not a small challenge.
- *Non-governmental organizations (NGOs)* – Many global, regional and local NGOs are at the center of addressing access to clean water and sanitation, influencing public policy, developing tools to promote sustainable water management (including water footprinting and water risk mapping), and working closely with civil society and the private sector. Innovation in partnerships with NGOs and the private sector are growing rapidly as there is recognition for "collective action" to address these water challenges. Many major multinational companies with water stewardship programs have

partnered with NGOs to tackle complex water issues. These NGOs are focused on building water sector technical capabilities within the public sector, and advocating for changes in public policy and water pricing to adjust to the new realities of water scarcity.

- *Civil society* – Almost everyone has an opinion on water and with the advent of social media they can broadcast that opinion far and wide with lightning speed. Now *everyone is a stakeholder* and, as a result, can weigh in on issues such as access (or lack of access) to clean water and sanitation and how both the private and public sectors are managing water. Civil society can influence the private sector on issues such as water use and as a result can put corporate reputations at risk. There are several widely publicized cases where civil society had the ability to withdraw a company's social license to operate when it was perceived that water was not being used wisely.
- *Governments* – Governments are challenged by a wide array of issues related to water: how to fund water infrastructure projects (capital investment and repairs); how to promote economic development in the face of increased droughts and water scarcity; and perceptions by civil society on the effectiveness of governments to deliver basic services such as access to clean water and sanitation. For various governments such as Singapore and Israel, water is a critical resource and water tech is a significant technology export opportunity. Meeting the public needs and driving economic benefit through technology export is a smart approach from any perspective.
- *Innovator/entrepreneur* – These are the people with the ideas and capital to meet a critical need such as water for agricultural, industrial and domestic use. Academics, inventors and investors are addressing several key issues and looking for capital to translate the ideas into commercial and scalable solutions.

The dilemma of water pricing

The price of water is typically insignificant on a profit and loss statement so why should anyone care about its price? In particular, why would the CFO care?

Let's start with an overview of the current price of water.

According to the recent Circle of Blue survey of US water prices performed in 2012,[3] the price of water in the US has increased an average of 7.3 percent during the last year and 17.9 percent since 2010, when Circle of Blue began collecting pricing data. The median increase was 7.8 percent over the last year.

A couple of examples of water prices in the US: the price of water in New York City is $76.08 for a family of four using 150 gallons per day per person. For residents in Phoenix, Arizona the price of water is $64.37 per month for the same usage.

Circle of Blue began tracking water rates in 2010 for the same 30 US cities (the 20 largest in the nation, plus 10 regionally representative cities). From 2010 to 2011, the first year an annual comparison was possible, prices increased an average of 9.4 percent, with a median increase of 8.6 percent. The prices are based on "medium consumption," which is defined as a family of four using 378 liters (100 gallons) per person per day – roughly the national average for daily per capita domestic water use as calculated by the US Geological Survey.

While those residents that pay the bills may view these prices as "high," the prices do not reflect the true cost of water *or its true value.*

The price of water varies considerably depending on how water infrastructure projects are funded. In the American West there is federal investment in large federal projects such as the Central Arizona Project, connecting the Colorado River to Phoenix, Tucson and other Arizona cities. These federal projects can create an artificially low price for water.

In contrast, when capital water projects are funded directly by customers the price of water increases. Let's use the city of Santa Fe, New Mexico as an example. The city of Santa Fe has the highest overall water rates in the US based upon the Circle of Blue survey. Water in Santa Fe is scarce. As a result, the city is building the Buckman Direct Diversion to capture water from implementation of the San Juan – Chama Project. This non-federally funded project is estimated to cost $217 million and is being paid by the city and county of Santa Fe. Mostly new water consumers in Santa Fe pay this cost.[4]

The 2010 Black & Veatch survey of the "50 Largest Cities Water and Wastewater Rate Survey," a resource for water and wastewater utilities, highlights customer charges for water and sewer service for residential, industrial and commercial customers.[5]

A key finding of the survey is that water and wastewater bills for US residential use have increased at a steady rate since 2001, when Black & Veatch began producing the survey. Analysis of the 2010 survey results indicates the average annual increase in typical residential water bills is approximately 5.3 percent from 2001 through 2009, while the increase in typical residential sewer bills is approximately 5.5 percent.

While water is still priced low, a steady increase in price of about 5 percent starts to become meaningful.

What about global water prices?

For the most part, the same holds true that for global water prices. The actual price of water has increased and is buried under subsidies, taxes and sunk costs of municipal and regional water departments.[6] Water prices are sensitive to the social, physical, institutional, and political setting and as a result most often do not reflect the actual costs for water infrastructure and delivery.[7] In addition, water prices are increasing, unpredictably at times. This creates uncertainty for businesses.

As long as water prices are well below the actual costs, incentives for conservation are very limited. Moreover, as long as the true value of water is not understood and communicated to all stakeholders we will be left to continue to wastewater.

One might think that businesses would want an artificially low price for water.

This is not the case.

In a *Wall Street Journal* op-ed piece entitled "Putting a Price on Clean Water," the rationale for putting a price on water was clearly articulated.[8] The authors cite the troubling statistics on access to clean water and provide an example of how a successful water policy and pricing program works. The Third World Centre for Water Management estimates that the number of people who did not have access to safe drinking water was at least 1.8 billion in 2009.[9] The Centre believes that "in most developing countries, it should be possible to provide clean drinking water to all urban centers of more than 200,000 people." The knowledge, funds, technology and experience exist to make this possible and it isn't happening because of "poor water management and governance practices, and the lack of political will."[10]

In contrast, they cite the example of Phnom Penh, Cambodia as effective water management – the practice of sensible water pricing as a means to ensuring clean water to all its citizens on a continuous basis. According to the authors, the Phnom Penh Water Supply Authority was nearly bankrupt and could provide poor quality water to only a small fraction of its population for two to three hours each day in 1993. And by the date of the op-ed piece, the city now provides clean water from domestic taps. The practice of equitable water pricing has resulted in a reduction of 70 to 80 percent in water bills and residents now receive clean water 24 hours a day in their homes in the city's poorest households in the slums.

This story of Phnom Penh highlights that effective and fair pricing of water can deliver clean water to the entire population, in particular the poorest of the city. If it is possible here, it should be possible for all.

Over the past few years there has been an increased focus on developing an understanding of the real value of water. The approaches range from a qualitative survey of how stakeholders view and value water to more quantitative approaches. Both perspectives provide valuable insight as to how stakeholders value water.

In tension with water prices that reflect the actual costs of water is the "human right to water" touched upon by Brabeck and colleagues.[11] The 1977 United Nations Water Conference in Mar del Plata, Argentina, established the concept of basic water requirements to meet fundamental human needs. This was reiterated at the 1992 Earth Summit in Rio de Janeiro, Brazil. General Comment 15 on the right to water, adopted in November 2002 by the Committee on Economic, Social and Cultural Rights, sets the criteria for the full enjoyment of the right to water.

However, in 2008 the UN General Assembly stopped short of declaring water as a human right. The UN Committee on Economic, Social and Cultural Rights states a human right to water in Articles 11 and 12 of the International Covenant on Economic, Social and Cultural Rights.[12] The covenant includes an obligation to prevent corporate third parties from infringing on the right to water by "polluting and inequitably extracting from water resources."

The CEO Water Mandate has weighed in on this topic with a discussion paper.[13] This was based upon a paper called *DRAFT Business, Human Rights, and the Right to Water*.[14]

The authors of the CEO Water Mandate discussion paper discuss the belief that water is a human right and they explore its implications for businesses. Specifically, the paper poses the question as to what would adoption of the "corporate responsibility to respect" principle look like?

The authors argue that a rights-based approach to water means that access to water for public use should be prioritized over other water uses, such as agriculture and industry, to maintain sufficient water supply for domestic use.

How do the public and industry view value of water and pricing?

The August 2009 Circle of Blue/Globescan survey provides insight as to how stakeholders view and value water.[15] The survey was conducted in 15 countries, and included surveys of 1,000 adults with "deep dives" into seven countries with 500 adult interviews per country.

A couple of key questions from the survey shed light on how stakeholders view water. When asked about the seriousness of water problems, 72 percent are "somewhat concerned" about water pollution and 71 percent are "somewhat concerned" about a shortage of freshwater. When stakeholders were asked about how concerned they were about the lack of safe drinking water about 84 percent of stakeholders in Mexico were very concerned. With India (67 percent), Canada (65 percent), UK (61 percent), China (59 percent), US (56 percent) and Russia (42 percent) following.

The 2012 report by Xylem on the value of water also provides insight as to how water is valued in the US.[16] The 2012 Xylem value of water index is based on a telephone study of 1,008 American voters age 18 years and older, and an oversample of 250 New York City residents. A few key findings from the Xylem study:

- 77 percent of Americans are concerned about the state of the US water infrastructure system.
- 88 percent of Americans believe the US water infrastructure needs some sort of reform, compared with 80 percent in 2010.
- The vast majority of Americans (88 percent) believe that government should be accountable for fixing and maintaining the nation's water infrastructure. They want government to invest more time (79 percent) and money (85 percent) in upgrading US water pipes and systems. Moreover, most Americans trust local and municipal governments to address these problems more than other entities.
- The amount Americans are willing to pay has increased by 24 percent since 2010.
- Despite a majority of Americans saying they are concerned about the state of the country's water infrastructure and are willing to pay more for improvements, there is a disconnect in understanding the issues. Americans are largely unaware of their water footprint or the extent to which water infrastructure problems would impact them personally. More than half of Americans believe they use 50 gallons or less daily while usage is closer to twice that amount.

• 29 percent of Americans believe that water infrastructure problems would affect them "a great deal," compared with 41 percent in 2010.

The intersection of how people value water and what they are willing to pay for infrastructure improvements to address access to clean water highlights the disconnect between price and value. How does the public sector create "sustainable infrastructure" to address water needs in view of this disconnect between price and value?

A recent report by Ceres in collaboration with American Rivers and the Johnson Foundation at Wingspread, titled *Charting New Waters, Financing Sustainable Water Infrastructures*,[17] also supports findings from the 2012 Xylem report. While focused on US water infrastructure it provides insight as to the challenges faced by the public sector in funding water infrastructure in a sustainable manner.

The key questions posed by the researchers were:

• What new financing techniques can communities pay for integrated and sustainable infrastructure approaches?
• How can we direct private capital toward more sustainable water management?

The report concludes: "While options for more cost-effective, resilient and environmentally sustainable systems are available, they are not the norm." In contrast, investment in inflexible and expensive "siloed" water systems is still pervasive, despite the fact that money available for financing water infrastructure is increasingly scarce." Also, these existing systems are wildly inefficient (losing an estimated 6 billion gallons per day of expensively treated water to leaking and aging pipes – about 14 percent of daily water use).

Key findings and recommendations from the Ceres report related to water pricing and infrastructure are paraphrased below.

• Recognize that local pressures will drive local solutions – water systems are as diverse as the drivers of change that impact them but solutions are emerging at the local level, including green infrastructure, closed loop systems and recycling. Financing models need to be developed that can support this type of local activity, which can then be scaled up.
• Consumers should be given choices and options – water systems typically provide one product at a single price, focusing on potable water. While that has worked well in the past, potable water is the most expensive kind of water and is widely used for non-drinking purposes such as watering lawns, flushing toilettes and showering. Consumers should be given options that include differentiated rates for drinking water versus other types. Additionally, water systems should explore how to move beyond "minimum cost rates" to meet customer demands.

- The financial health of US water systems is directly linked to their long-term sustainability – US water systems need to embrace various financing changes in order to ensure long-term sustainability. These include full-cost accounting of water services; incorporating value-added services into the revenue picture to better align customers' perceived value with products delivered; improving the capture and dissemination of performance data to drive efficiency; and considering consolidation of certain systems to enhance efficiency.
- Innovative financing models should be pursued to increase efficiency, add value to customers, and lower costs for providers – models should include: mechanisms to expand the pool of water service funding to non-traditional partners; increasing incentives and markets for distributed water services that include "low-impact development," such as on-site treated wastewater for buildings; and other green infrastructure initiatives.
- Alternative market-based solutions should be explored and evaluated for scalability – solutions could include: properly valuing and pricing ecosystems services, which provide enormous value yet are largely unaccounted for in the present system; developing securities to aggregate customer-financed projects such as greater "where it falls" water management; and creating private investment opportunities for efficiency gains from such things as retrofitting and closed-looped water systems to reduce system impacts and improve efficiency at both the building and neighborhood levels.

If current (or unsustainable) water pricing frameworks do not reflect the actual costs for water, what does "sustainable water pricing look like? Can water be priced according to its value while acknowledging the human right to water? This is the challenge of sustainable water pricing. Sustainable water pricing will require a system-based approach and integration of factors such as equity in water availability and affordability.[18]

Beecher and Shanaghan maintain that sustainable water management requires both appropriate price signals and a balance between other policy goals such as affordability and equity among stakeholders. Moreover, they recommend that a sustainable water price is one that will:

- reflect true costs and induce efficient water production and consumption;
- promote optimization or the achievement of least-cost solutions to providing water service;
- achieve equity in terms of incorporating cost-sharing practices as needed to enhance affordability; and
- enhance the long-term viability of the water utility.[19]

They also make a key point; in establishing sustainable water pricing the size of the system is important. The larger water systems (as dictated by the boundaries of the water service area and utility) are more likely to achieve optimal pricing solutions as they can spread the cost of service in an equitable manner for the customer base. Moreover, they maintain that sustainable water pricing may require an "evolution from the somewhat rigid doctrine that guides pricing today" such as marginal-cost pricing principles.

What is the business value of water?

Before we tackle the business value of water let's further examine the cost and price of water.

- Cost is defined as the expense of producing and delivering a unit of a commodity or resource, environmentally and technologically determined. Costs are incurred at each stage of the value chain and can be monetary or non-monetary. Some of these costs are internalized in the product price, but others are not (externalities).
- Price is the monetary expense to consumers of purchasing a unit of a product or service. This is highly variable based on the environmental, economic, political and social setting of the market transaction.
- Value is the worth of something in a given context. It is intrinsically social, as valuation processes involve ethical and moral choices about personal and societal priorities. Value can be material, monetary or non- monetary, and can be articulated through the application of different valuation methods (e.g. economics).

Now what is the value of water to business?

The Initiative for Global Environmental Leadership (IGEL) at Wharton, University of Pennsylvania released a report addressing the business value of water, *Valuing Water: How can Business Manage the Coming Scarcity?*[20] The IGEL report captures the risks to businesses from water scarcity along with the real value of water from a business perspective.

The real value of water is tied to issues such as business continuity (which includes social license to operate) and reputational risk/brand as a driver of innovation. For now, the real value of water for a business is mostly tied to business intangibles (license to operate, reputation/brand value and innovation) as long as the price of water is an insignificant operating cost. Let's examine these key aspects of value – business continuity, social license to operate, reputational risk/brand value, and innovation.

Business continuity

Water scarcity can represent a physical risk to a business. Essentially this means that a business operation can be disrupted if there is no water to run production processes, for product ingredients and as components of their supply chains.

Moreover, a business is disrupted if water is not available for consumers to use the products manufactured by a company.

Water scarcity, which disrupts a business, can be quantified in economic terms. It is the economic value lost, the difference between what happened and what would have happened except for the disruption. While the concept is simple, the calculations can be complex.

How does a business minimize the risk of disruption and ensure continuity? If water is a key component of your upstream or downstream value chain and /or used in the manufacturing processes and product ingredient, then it is critical that water is managed in a manner to ensure business continuity in the event that water is not available or supplies are interrupted.

License to operate

License to operate has become a critical business risk to be managed. No longer is license to operate a matter of business licenses and permits. License to operate can be given and withdrawn by a variety of stakeholders important to a business. Nowhere has this been more important and visible than in developing economies in Africa and India. Global food and beverage companies know all too well the risk of not engaging stakeholders on critical resource issues such as water. The emerging issue of the human right to water will only make the competition for water more acute. This competition, in turn, increases the risk to a company's license to operate unless water stewardship is a key aspect of business strategy, governance and operations for both direct and indirect water use.

Reputation and brand value

There is much discussion of reputation and brand value these days as the notion of a "green" brand has entered our lexicon. A *brand is a promise* and a relevant and distinctive promise helps to build a brand. A corporate *reputation is built by fulfilling that promise* to stakeholders. Simply put a *company owns its brand, but stakeholders own its reputation*. If a company breaks a promise to stakeholders, its reputation and, in turn, its brand can be damaged. A promise such as operating in a sustainable manner with respect to water use has the ability to enhance reputation and brand value.

Innovation value

Sustainability is a driver for innovation and water scarcity will drive innovation in new products and services. The rest of this book will expand on this key issue of "value."

A great way to view the value of water is to note the increased attention coming from the financial sector and industry analysts. A 2011 report by Morgan Stanley titled *Peak Water: The Preeminent 21st Century Commodity Story*[21] provides a well-rounded view of the importance of water and how investment is moving into water tech to address a number of key issues.

Box 1.1 Water risk = money

In *Watching Water: A Guide to Evaluating Corporate Risks in a Thirsty World*, JP Morgan analysts begin by stating that:

> A scarcity of clean, fresh water presents increasing risks to companies in many countries and many economic sectors. These risks are difficult for investors to assess, due both to poor information about the underlying supply conditions and to fragmentary or inadequate reporting by individual companies. As a result, market prices of securities are unlikely to accurately reflect the potential costs of water-related problems.[22]

These are profound comments.

The US Securities and Exchange Commission even notes in interpretive guidance for companies that "changes in the availability or quality of water...can have material effects on companies."[23]

Corporate disclosure of material business issues is a core foundation for smart investment decision-making, according to a benchmark study on water risk, *Murky Waters? Corporate Reporting on Water Risk* by the Boston, Massachusetts-based NGO group, Ceres. The report says:

> Emerging risks and opportunities that will impact corporate bottom-lines – including those associated with environmental, social, and governance (ESG) issues – must be included in financial filings.
>
> Global water scarcity is one emerging risk that all companies should be focused on – and one about which investors need information. The combination of rising global populations, rapid economic growth in developing countries, and climate change is triggering enormous water availability challenges around the world.
>
> We're already seeing tangible impacts from this issue. In the past two years, water shortages in California have shuttered new housing developments and forced farmers to abandon or leave unplanted more than 100,000 acres of agricultural land, resulting in more than $1 billion in lost revenue.[24]

Therefore, water risk is increasingly a potential financial risk.

The key issues identified by the recent Morgan Stanley report[25] include: increasing water scarcity; the relationship between water and food security; the energy–water nexus; water and shale gas development; and the impact of water scarcity on China's economic development. All of these issues are important to businesses.

According to Morgan Stanley, the "global water industry is expected to undergo a substantial transformation in the near future. Businesses will need to make further investments in water technology, and utilities will need to devote more money to water infrastructure."[26] They estimate that capital expenditures on water infrastructure are expected to grow to $131 billion in 2016 from $90 billion in 2010 (according to Global Water Intelligence (GWI)) with sales of water- and wastewater-treatment equipment to industrial users expected to rise to $22 billion by 2016, up from $14 billion in 2010, a 7.8 percent compound annual growth rate.[27]

Morgan Stanley expects that changing financial models in the municipal water sector will drive much of this investment growth. Traditionally, less than half of the money invested by utilities is derived from their own operations, with the rest of the costs being paid by local governments. Given the increasing financial pressure on the public sector, this model is no longer sustainable. Again, the issue of water pricing and value is central to driving investment in the water industry.

Industry analysts are also weighing in on the water industry and the value of water to businesses. Verdantix published a report titled *The State of Corporate Water Strategies*, which closely tracked the results of the 2011 CDP Water Program.[28]

According to Verdantix, over the past two years water management has become an "increasingly visible issue within corporate sustainability." Verdantix surveyed 100 senior executives in firms generating more than $1 billion in revenues to capture the organizational maturity of water strategies in both water-intensive and non-water-intensive firms. Verdantix heard that in 2011, most firms perceive water issues as short-term risks, concentrating their efforts on compliance strategies. At the same time, water stewardship, characterized by firms working beyond compliance requirements, is an immature area.[29]

Looking forwards, Verdantix found that most firms aim to realize tangible financial savings from their water strategies, while a leading minority of firms seeks to distinguish themselves by targeting stewardship opportunities.

So, the key question becomes: if water does not reflect its full price let alone full value, how can we drive innovation in the water industry? We will provide some of the answers in the following chapters.

Notes

1   Ernst & Young with GreenBiz Group (2012) *Six Growing Trends in Corporate Sustainability*.
2   www.unwater.org/downloads/Water_facts_and_trends.pdf, page 3.
3   www.circleofblue.org/waternews/2012/world/the-price-of-water-2012-18-percent-rise-since-2010-7-percent-over-last-year-in-30-major-u-s-cities, 10 May 2012.
4   www.circleofblue.org.
5   www.wichita.gov/NR/rdonlyres/397682AD-954B-4422-9C56-D9D121F19B88/0/200750LargestCitiesSurvey.pdf.

6   www.euractiv.com/cap/business-frets-growing-water-gap-news-223040.
7   http://elibrary.worldbank.org/content/workingpaper/10.1596/1813-9450-2449.
8   Peter Brabeck, Asit K. Biswas and Lee Kuan (2011) "Putting a Price on Clean Water," *Wall Street Journal*, www.wsj.com, March 21.
9   www.thirdworldcentre.org.
10 Ibid.
11 Brabeck *et al.*, *op. cit.*, note 8.
12 WHO (2007) *Health and Human Rights: International Covenant on Economic, Social and Cultural Rights*, www.who.int/hhr/Economic_social_cultural.pdf (accessed 30 June 2010).
13 Tripathi, S. & Morrison, J. (2009) *Water and Human Rights: Exploring the Roles and Responsibilities of Business*, The CEO Water Mandate Discussion Paper, March, www.unglobalcompact.org/docs/issues_doc/Environment/ceo_water_mandate/Business_Water_and_Human_Rights_Discussion_Paper.pdf (accessed 30 June 2010).
14 Institute for Human Rights and Business (2009) *DRAFT Business, Human Rights, and the Right to Water: Challenges, Dilemmas, & Opportunities Roundtable Consultative Report*, January, www.ihrb.org/publications/reports (accessed 4 December 2012).
15 www.circleofblue.org/waternews/2009/world/waterviews-water-tops-climate-change-as-global-priority.
16 Xyleminc (2012) *Value of Water Index – Americans on the US Water Crisis*, www.xyleminc.com.
17 Ceres, American Rivers and the Johnson Foundation (2011) *Charting New Waters, Financing Sustainable Water Infrastructures*, July–August, www.ceres.org, www.americanrivers.org, www.johnsonfdn.org.
18 Beecher, J. A. and Shanaghan, P. E. (2001) *Sustainable Water Pricing*, http://opensiuc.lib.siu.edu/jcwre/vol114/iss1/4 (accessed 4 December 2012).
19 Ibid.
20 http://igel.wharton.upenn.edu, March 2011.
21 Morgan Stanley (2011) *Peak Water: The Preeminent 21st Century Commodity Story*, November.
22 JP Morgan Global Equity Research (2008) *Watching Water: A Guide To Evaluating Corporate Risks in a Thirsty World*, April.
23 www.sec.gov/rules/interp/2010/33-9106.pdf, page 6.
24 www.waterfootprint.org/Reports/Barton_2010.pdf.
25 Morgan Stanley, *op. cit.*, note 21.
26 Ibid.
27 Ibid., figure 3, page 3.
28 Verdantix, *The State of Corporate Water Strategies*, www.verdantix.com.
29 Ibid.

Chapter 2

Global trends as drivers for innovation

What drives innovation and what drives innovation in the water industry?

What are the challenges to innovation in the water industry and how can these challenges be overcome?

The forces at work in driving innovation in the water industry are global. In addition, over the past several years the sustainability movement has been increasing in momentum, resulting in new products and services unimagined a short time ago.

Insight into these global trends and sustainability will set the stage for understanding water tech innovation.

Global trends as drivers for innovation

Water scarcity is when water demand exceeds supply, which is further complicated by the impacts of droughts and climate change. As briefly discussed in Chapter 1, the reasons for water demand exceeding water supply are provided in a publication by JP Morgan Global Equity Research.[1] The JP Morgan study identified the risks of current and projected water scarcity to businesses and investors. A brief summary is provided below.

- Water is increasingly scarce due to population growth, urbanization and the impacts of climate change.
- Investments are available in areas such as water supply infrastructure, wastewater treatment and water demand management.
- Water scarcity risks vary between industry sectors and geographic locations due to climatic conditions, water resources and water use practices.
- The financial impact of water scarcity on specific industry sectors and companies is unknown as information regarding water use is incomplete.

According to the JP Morgan report, the principal factors resulting in "imbalance" between water supply and demand are outlined below.

- *Population growth and increasing food needs (the rise of the middle class).* The current global population recently crossed 7 billion (at the time of the JP

Morgan report it was about 6.4 billion) and is increasing at about 70 million per year, with most of the growth in emerging economies. The global population is expected to reach 8.1 billion by 2030 and 8.9 billion by 2050. While growth in Organisation for Economic Co-operation and Development (OECD) countries is expected to remain relatively flat, the population of the US is expected to increase to 370 million by 2030. In general, as population increases so do water withdrawals.

- *Urbanization.* Greater than half of the global population now lives in cities and increasing urbanization results in increased industrialization and increased water use.
- *Climate change.* Climate change will alter hydrologic cycles on both a regional and local level. The long-term and short-term availability of freshwater will be altered along with changes in water quality (water temperature, dissolved constituents, etc.).[2]

Observations by JP Morgan on scenarios for water supply and demand indicate "worsening trends" and increasing threat to industry sectors and the public. Of particular note is the impact of water scarcity at the local (watershed level). Their conclusions are summarized below.

- *As available water supply becomes less reliable, data also suggests worsening trends in water quality in certain regions.* In developing countries water quality is deteriorating. In particular, fast-paced (growth outpacing infrastructure support) urbanization increases stress on water quality through releases of untreated sewage and wastes.
- *Water-quality issues interact with availability concerns.* Water supply and water quality are closely connected. Examples include how excessive groundwater pumping can result in saltwater intrusion, where saltwater is drawn into freshwater aquifers (along coastal areas) and where releases of contaminants into groundwater limit the amount of clean groundwater available for potable use.
- *Water-quality data are location specific.* Water quality data such as biological oxygen demand and nitrogen loading reflect the unique aspects of a watershed.
- *Businesses face three types of water-related risks: physical, regulatory and reputational.* Physical or operational risk mostly impacts industry sectors in which water is consumed or evaporated in the production process (e.g. the utility sector). A lack of water of adequate quality directly reduces production in these sectors; agriculture, beverages and food processing. Regulatory risk is most critical for industry sectors that use or discharge relatively large amounts of water in connection with relatively low-value production processes (such as the energy sector). Finally, reputational risk is tied to increased competition from economic, social, and environmental interests. This risk has a significant potential to damage a company's reputation and

limit their license to operate (such as the food and beverage sectors). Multinational companies operating in developing countries are especially vulnerable to reputation risk from water.

- *These three types of risks are closely related and manifest in combination.* For example, water scarcity (physical) may lead to the revocation of water licenses (regulatory), or to damage to a firm's image and brand (reputation). Physical, regulatory and reputation risks may impact at different points along the value chain and may affect suppliers, production facilities, or users of the product.
- *Companies may have significant exposure to water scarcity and water quality even in countries where they do not have production operations.* Backward linkages (supply chain) and forward linkages (product use) may create water risks that go unnoticed by management and investors.
- *Many industry sectors have a larger water footprint in their supply chain than in their direct production.* An example is the food and beverage sector, which relies upon irrigated agriculture for critical inputs. As a result, water scarcity in key production areas could lead to higher prices for grain, meat, and other inputs. Also, aluminum manufacturers in the northwest US experienced supply-chain impacts in 2001, when water shortages led to the curtailment of hydroelectric power supplies, forcing the closure of several aluminum plants.
- *Product use can be problematic.* The use of consumer products such as washing machines requires water for use, and a scarcity of water will impact their use.[3]

Water risk will potentially impact the financial performance of companies in several ways. Financial performance can be impacted by water-related issues through the disruption of production, supply chain disruption, higher costs in the supply chain (especially in the agricultural sector), changes in production methods (from improved water efficiency), capital expenditures to change production processes (to improve water efficiency) and to secure, recycle or treat water, increased regulatory compliance, and increased costs to procure and discharge water. In addition there are impacts to limits of growth and costs to reputation and brand (intangible value).

A 2030 Water Resources Group (WRG) report[4] provided a quantitative view of supply and demand projections. WRG consists of stakeholders who are concerned about water scarcity "as an increasing business risk, a major economic threat that cannot be ignored, and a global priority that affects human well being."

WRG concluded that "there is little indication that, left to its own devices, the water sector will come to a sustainable, cost effective solution to meet the growing water requirements implied by economic and population growth."[5] This is not an encouraging prognosis for the future considering the increasing demand for water by both the public and private sectors.

WRG makes the key points that, in "the world of water resources":

- economic data are insufficient;
- management is often opaque; and
- stakeholders are insufficiently linked.

WRG lays out plausible scenarios for water supply, water demand and the "water gap" on a regional (local) scale. This gap will play a critical role in how businesses address the risk and opportunities for their businesses. WRG concludes that by 2030, assuming an average growth scenario and if no efficiency gains are realized, global water requirements will grow from 4,500 billion cubic meters to 6,900 billion cubic meters. As illustrated below this is about 40 percent above current accessible and reliable supplies.

The 40 percent gap is driven by global economic growth and development. Agriculture makes up the majority of this global water demand with current use at 3,100 billion cubic meters or about 71 percent of total demand. This is expected to increase to 4,500 billion cubic meters by 2030, which, given projected population growth, is a slight decline to about 65 percent of total water demand. Industrial demand is currently 16 percent, with a projected increase to 22 percent by 2030. Domestic water demand will decrease slightly from 14 percent to 12 percent by 2030.[6]

The question posed by WRG is how this gap will be closed. An examination of the projected gap (business as usual) and estimated increases in supply and water efficiency was examined. For example, if one looks at historical improvements in water efficiency in agriculture it is only about 1 percent improvement between 1990 and 2004. A similar rate of improvement was observed in the industrial sector. If these rates of efficiency improvements are projected to 2030 we would only meet about 20 percent of this 40 percent gap. If we assume a 20 percent increase in supply we would still have a remaining 60 percent of demand unmet.

A few of the conclusions outlined by WRG are as follows.

- *Agricultural productivity is a fundamental part of the solution to close the water gap.* Focusing on agricultural productivity is essential in closing the water gap since the agricultural sector contributes the greatest to global water use and water efficiency is one of the key low cost technology solutions. Increasing the "crop per drop" can be accomplished through an integrated approach of increased drip and sprinkler irrigation no-till farming and improved drainage, drought resistant seeds, optimizing fertilizer use, crop stress management, improved best practices (integrated pest management) and innovative crop protection technologies. For example, WRG analysis of India indicates that low-cost technologies applied in the agricultural sector can close about 80 percent of the water gap.
- *Industrial and municipal productivity is just as critical as agricultural productivity improvements.* China provides an excellent example as to how industry

and municipal water efficiency projects can provide a positive contribution to closing the water gap. Although agriculture is about 50 percent of water use in China, the fastest water use is occurring in the industrial sector. Water efficiency programs, in particular with new construction, could close the water gap in China by about 25 percent with a savings of about $24 billion.

- *Quality and quantity of water is linked*. WRG evaluation of São Paulo, Brazil provides an example of how water quality and water quantity issues and solutions are linked (there really is no separation). The recommended approach to closing the water gap in Sao Paulo is in water efficiency and productivity gains. Water quality improvements are also a critical aspect of overall water management issues. Financial benefits to the industrial sector can be obtained from reduced water use and municipalities can save 300 million cubic meters of water through utility leakage reduction. Wastewater reuse for gray water purposes (such as industrial processes and public works uses) can save about 80 million cubic meters in "new water" requirements.
- *Most solutions require cross-sector tradeoffs*. South Africa provides an excellent example of a balanced approach to water stewardship. Cost-effective measures are applied in water supply (water supply increases can close the water gap by about 50 percent), agricultural efficiency, and productivity improvements (about 30 percent of the gap) and industrial and domestic solutions (the remaining 20 percent of the gap). As one would expect, the watersheds dominated by agricultural use rely on agricultural improvements and industrialized watersheds (Cape Town and Johannesburg) rely on industrial and domestic solutions.

It is projected that by 2025 approximately 47 percent of the world's population will face water scarcity.[7] This, coupled with a focus on the "human right to water" voiced by the UN, will increase competition for water resources. As competition increases so will technology innovation to increase supply and better manage demand. This innovation will be driven by the private sector as it moves to ensure adequate water for operations, preserving license to operate and brand/reputational value, and seeks to develop new products and services.

The role of reporting and disclosure in innovation

Increased reporting of water-related risks and opportunities is also driving innovation. One key reporting framework is the CDP Water Program. Since 2003 CDP has grown to over 3,000 organizations that quantify and disclose their greenhouse gas emissions, water management and climate change strategies. These organizations are setting reduction targets and advancing resource improvements. The data from CDP are made available for use by institutional investors, corporations, policymakers and their advisors, public sector organizations, government bodies, academics and the public. In late 2011 CDP represented 551 institutional investors, holding US$71 trillion in assets under management.[8]

Box 2.1 What is the Carbon Disclosure Project?

The Carbon Disclosure Project (CDP) works to transform the way the world does business to prevent climate change and address water-related issues (CDP Water Program). The CDP says it believes in a world where capital is efficiently allocated to create long-term prosperity rather than short-term gain at the expense of the environment.

The CDP utilizes measurement and information disclosure to improve the management of environmental risk. It does this by leveraging market forces, including shareholders, customers and governments. In this way the CDP has incentivized thousands of companies and cities across the world's largest economies to measure and disclose their greenhouse gas emissions, climate change risk and water strategies.

The CDP, based in London, holds the largest collection globally of self-reported climate change data. With this knowledge companies and cities are better able to mitigate risk, capitalize on opportunities and make investment decisions that drive action towards a more sustainable world.

It also works with 50 purchasing organizations engaged with CDP to mitigate environmental risk in their supply chains.

In 2010, CDP launched the CDP Water Program, and is now taking the same approach to identifying water risk and opportunities as it did with climate change by using reporting and disclosure to move businesses to develop strategies to mitigate water-related risks. Each year since 2010, CDP Water Program has published reports providing detailed analysis of the responses indicating important trends and developments.

CDP is aware that knowledge and awareness drives action, accountability and innovation.

In 2009, CDP launched the Water Program, with the first report in 2010. The results of the 2011 CDP Water Program provide insight as to how companies view water related risks and opportunities.[9]

A summary of the 2012 CDP Water Program for the Global 500 companies[10] is provided below.

- More than half of Global 500 respondents (53 percent) have experienced detrimental water-related business impacts, such as business interruption and property damage from flooding, with associated financial costs for some companies as high as US$200 million; this figure is up from 38 percent last year. Perhaps as a result, more respondents (68 percent) report exposure to water-related risks, up from 59 percent.
- Water represents a strategic opportunity to improve financial and brand performance: 71 percent of respondents reported a total of 319 water-related opportunities such as the sale of new products or services; 79 percent of

opportunities reported with an associated timeframe are expected to materialize now or within the next five years, some with a sales potential of more than €800 million by 2020.

- Water is still not receiving the boardroom attention it deserves The proportion of respondents with board-level oversight of their water-related policies, strategies, or plans is essentially unchanged from 2011 at 58 percent. Furthermore, the proportion of respondents setting concrete water-related goals and targets has also changed little at 55 percent.
- Assessing and addressing exposure to water-related supply chain risk is on the rise. There has been a marked increase in awareness of supply chain risks with 71 percent of respondents now able to state whether or not they are exposed to such risk (up from 62 percent in 2011). However, 29 percent of respondents remain unaware. Similarly, more respondents (39 percent) are now requiring their key suppliers to report on water-related risks than ever before (up from 26 percent in 2011), although there remains plenty of room for improvement.
- Companies traditionally work independently to tackle water-related issues. While these initiatives drive efficiency or quality improvements, they are often limited in scope. Given the complexity and scale of water challenges and the interdependencies between companies, communities and natural ecosystems, stand-alone actions may no longer be enough. Global 500 companies realize that working collectively with a range of partners, beyond the boundaries of their direct operations, can effectively build resilience and add value across their business as well as for the other users of this shared resource.
- 74 percent of respondents report taking some form of collective action to address water-related issues with benefits ranging from increased business continuity, securing licenses to operate and increased brand value alongside the opportunity to gain fresh ideas, increase the momentum for change and pool resources.
- This year's respondents indicate that collective action will continue to be featured prominently in companies' water strategies and, in turn, overall business strategies. As the strategic importance of water-related issues grows it is anticipated that more Global 500 companies will leverage collective action in response.

What the 2012 CDP Water Program report tells is that water-related risks are near-term (in many instances being felt now by businesses) and that companies recognize that there are opportunities in new technologies, improved operating efficiency and improved brand value.

Sustainability as a driver for innovation

Over the past several years it has been effectively argued that sustainability is driving innovation. It is increasingly clear that this is the case as new technologies and services emerge to address a wide range of resource issues such as energy, carbon, water and material use. Let us examine the linkage between sustainability and innovation, as it will provide a better understanding of how innovation is reshaping the water industry.

The place to start in understanding the linkage between sustainability and innovation is to look at the work of Joseph Schumpeter.[11] Schumpeter believed that "waves" of innovation occur through new technological advances essentially "destroying" companies and industries that can't adapt and creating new industries and businesses.

Described as the dynamic between start-up companies and established companies is a process he referred to as "creative destruction." This process is not continuous improvement but instead the emergence and embracing of disruptive technologies.

It was Hart and Milstein in *Global Sustainability and the Creative Destruction of Industries* who linked sustainability to creative destruction.[12] Hart and Milstein's thesis is that episodes of creative destruction are driven by waves of scientific and technological discovery and we are in the early stages of creative destruction as sustainability takes hold in the private and public sectors. The reasoning is that historically businesses had access to abundant raw materials (such as water), cheap energy and limitless sinks for waste.

With regard to water, we are moving from access to unlimited cheap supplies to increased water scarcity – increased competition for a finite resource. This scarcity impacts most businesses but some more acutely than others. Resource-intensive industries will find increasing water constraints a challenge and an opportunity for potential repositioning and new competency development.

Schumpeter was skeptical of the ability of businesses with large installed assets to make their established positions obsolete through innovations. However, he did recognize that large companies have the financial, technical and organizational resources typically unmatched by small entrepreneurial companies. When one views the landscape of water tech innovation, we see innovation coming from both large companies and start-ups (more in Part II of this volume).

We are seeing the progress large companies are making in addressing resource constraints by embracing sustainability and proactively managing resources such as water. This ability to improve water efficiency, recycling and reuse, and secure additional sources of water, are providing companies with better business resiliency, license to operate, and improved brand value.

Hart and Milstein cite the transformation of companies within the chemical sector as an early sign of the impact of sustainability on innovation. The move by a global chemical company to spin off their chemical businesses to focus on biotechnology is such an example. However, since their paper was first presented in 1999, there have been numerous examples of how industries are being transformed. More recently, examples can be found in the consumer packaged goods (CPG) sector (the move to concentrated detergents and cleaning products), the automotive sector (the move to hybrid and electric vehicles and the emergence of startup companies) and the energy sector (the adoption of biofuels and renewable technologies).

The CPG sector has some of the greatest resource challenges and opportunities as CPG products can have relatively "large" resource footprints. As a result there are opportunities for creative destruction to reduce these footprints both enterprise wide and product level. Some of the key questions being asked by companies are:

- How can we sustain the manufacturing of products with less use of resources (or in some cases entirely eliminate resources)?
- Can we grow or move into emerging markets when resources are constrained or too costly to provide any profit margin?

Within the CPG sector, food and beverage companies are leading in developing water stewardship strategies and driving innovation in water efficiency, recycling, reuse and conservation. As an illustration, for food and beverage companies to operate in water-stressed areas there will be a need to re-examine how water is used and managed with other shareholders within a watershed. Conversely, entrepreneurs will view water constraints as an opportunity to create new technologies in water efficiency, reuse and recycling, and so on.

The ideas of Hart and Milstein were built upon in an article titled "Why Sustainability is now the Key Driver of Innovation."[13] While the authors do not explore the roots of innovation, they do cite numerous examples of how companies are embedding sustainability into their strategies and developing new products and business lines.

The authors identify four stages of leveraging sustainability into a business strategy:

- Stage 1 – Viewing compliance as opportunity.
- Stage 2 – Making supply chains sustainable.
- Stage 3 – Designing sustainable products and services.
- Stage 4 – Developing new business models.

Opportunity for innovation resides in all four stages. However, Stages 2 and 3 represent particular opportunities for innovation, with Stage 2 as the area where opportunities for resource efficiencies, such as with water, reside.

What are the water tech opportunities?

What are the specific problems we are trying to solve? And what do the opportunities look like to meet the water needs of the public and private sectors, and also ecosystems?

Technology innovation will fall generally into addressing increasing supply, water treatment (desalination, reuse and recycling) and data acquisition/analytics that will positively impact supply and demand needs.

Overviews of the key focus areas to increase supply and effectively manage demand are outlined in two recent reports:[14]

- *Infrastructure*. According to the US Environmental Protection Agency (USEPA),[15] approximately $335 billion of investment is required over the next 20 years to maintain the US drinking water infrastructure at current levels. As an example, about 17 percent of treated water in the US is lost to leaky pipes; Boston loses about 30 percent of its water.

    There are no shortages of infrastructure technology opportunities but some of the more interesting are in infrastructure diagnostics, in-situ pipe rehabilitation, and water loss management.
- *Conservation and efficiency*. As with smart energy strategies, the place to start with water is with conservation and efficiency. Conservation and efficiency essentially creates additional water for use, and where conservation and efficiency are promoted, water use decreases. For example, in Albuquerque (New Mexico) and Las Vegas (Nevada), water use has declined from about 200 to 300 gallons per day to about 100 gallons per day. An essential component of increased conservation and efficiency is the pricing of water. Australia has focused on the best use of scarce water resources and as a result has moved to less water-intensive crops (from an exporter of rice to crops such as grapes and citrus fruits).
- *Treatment – recycling and reuse*. Expect to see no shortage of technology innovation opportunities as the need for water recycling and reuse increases to address water scarcity. There may also be a move to treat water at the appropriate level of use. For example, in the US only a small percentage of our drinking water is used for drinking. Of the approximately 130 gallons of water used per day that we treat to drinking water standards, most individuals drink less than a gallon per day. Including water to cook and clean with, this is still less than 15 percent of the total water treated. The rest is used to flush toilets, water lawns, and so on.

    Active treatment technologies such as activated carbon, ozonation, ultraviolet radiation disinfection, microfiltration and ultrafiltration, and membrane reactor technologies are all expected to grow. The implications for the industry growth rate for treatment chemicals is uncertain in view of new treatment technologies coming into play. Also, not to be ignored is the use of constructed wetlands which are passive treatment systems.

    Further driving innovation in the water treatment industry will be the emergence of new contaminants requiring treatment and lower regulatory requirements.
- *Interdependence between water and energy*. Currently there is much discussion of the energy–water nexus (and the energy–water–food nexus). To illustrate the importance of the linkage, in the US about 6 percent of energy consumption goes to moving water around and in California it is about 20

percent. Current projections of energy and water requirements lead one to conclude that business as usual cannot meet the resource needs of the future. As a result, there is and will continue to be an increased focus on energy-efficient water extraction, transport, and treatment technologies. Conversely, there will be an increased focus on energy generation with a lower water footprint – perhaps a move to renewable energy in part because of lower water requirements?

Box 2.2 The water–energy nexus

It's a simple equation: saving water saves energy, and saving energy saves water.

There is a growing awareness that water and energy issues are closely connected. What's not yet widely understood is just how much water we can save by saving energy, and how much energy we can save by saving water. The potential is enormous. With demand for water and energy continuing to grow, addressing the water–energy nexus is an opportunity we cannot afford to miss.

The River Network, a US non-profit, is dedicated to raising awareness about the connection between water and energy. Their position is:

> The more energy we save, the easier it is to reduce the harmful effects of our greenhouse gas emissions. The more water we save, the easier it is to secure precious freshwater resources and maintain a healthy, climate-resilient environment. Understanding these relationships between water and energy is more important than ever in today's changing world.[16]

Here's why that understanding is so important: the United States alone uses about half a trillion kilowatt hours per year of energy – 13 percent of the nation's total electricity use – to manage water. This is equal to the output of over 150 typical coal-fired power plants, the River Network notes:

> The bad news is that saving energy through water conservation, efficiency and reuse is not currently being utilized as a major strategy for addressing climate change. The good news is that this is one of the largest categories of energy use that we could reduce quickly and significantly, with the added benefit of protecting our water resources from the threat of a changing climate.[17]

"Smart water" and "smart energy" are inseparable.

- *Enhanced monitoring and measurement.* The saying, "what gets measured gets managed" is true in the water industry. Unfortunately, not much is measured, especially with regard to the quantity of water used or available resources (groundwater levels are a good example). There will be a move to real-time monitoring of water quantity use and quality. In addition, real time and wireless micro level soil moisture monitoring has the potential to improve agricultural productivity (this is significant, as about 70 percent of global water use is in the agricultural sector), save energy, reduce fertilizer usage and reduce waste and wastewater generation.

• *Sludge management and wastewater resource recovery.* There are several drivers and opportunities related to wastewater treatment sludge and wastewater. First, the demand for disposing of large volumes of wastewater treatment plant sludge is becoming challenging. Sludge treatment and disposal costs represent a significant portion of overall treatment costs. Couple this with expected increase in regulatory requirements and innovation solutions to sludge generation and management become attractive. There is also the recognition that waste sludge is a potential source of energy and resources/nutrients such as phosphates.

Other trends to watch that will drive innovation include urbanization and distributed water treatment. The majority of the world's population now lives in cities – we have become an urban world. With this comes a more energy intensive delivery of water services. In addition, the impact of sustainable land use and green building (the adoption of porous pavement and green roofs, for example) will impact (in a positive manner) how water is managed in the urban environment.

Also expect to see an increase in distributed water treatment systems. In the developed economies we have centralized water treatment systems that treat water to drinking water systems; we transport that water (with associated leakage) to users. The water is then used for a variety of purposes such as washing and sanitation. More tailored treatment and local/on-site treatment alternatives provide an option to reduce energy costs associated with treatment and transport (and a reduction in chemical usage in centralized systems).

It is clear that innovation in the water industry will come from public–private partnerships and investment in increasing water supply, managing demand, and leveraging IT to improve water data acquisition and analytics.

More on this in Part II of this book.

Notes

1   JP Morgan Global Equity Research (2008) *Watching Water: A Guide To Evaluating Corporate Risks in a Thirsty World*, April.
2   Ibid.
3   Ibid.
4   2030 Water Resources Group (2009) *Charting Our Water Future: Economic Frameworks to Inform Decision-Making*, www.2030waterresourcesgroup.com/water_full/Charting_Our_Water_Future_Final.pdf.
5   Ibid.
6   Ibid.
7   Ibid.
8   www.cdproject.net.
9   https://www.cdproject.net/water.

10 https://www.cdproject.net/CDPResults/CDP-Water-Disclosure-Global-Report-2012.pdf.
11 E.g. Joseph Schumpeter (1934) *The Theory of Economic Development*, Cambridge, MA: Harvard University Press; Joseph Schumpeter (1942) *Capitalism, Socialism and Democracy*, New York: Harper Torchbooks.
12 Stuart L. Hart and Mark B. Milstein (1999) *Global Sustainability and the Creative Destruction of Industries*, MIT Sloan, 15 October 15.
13 Ram Nidumolu, C. K. Prahalad and M. R. Rangaswami (2009) "Why Sustainability is Now the Key Driver of Innovation," *HBR*, September.
14 Steve Maxwell (2011) *2011 Water Market Review: A Concise Review of the Challenges and Opportunities in the World Water Market*, Techknowledgey Strategic Group; Sze Chai Kwok, Heather Lang and Paul O'Callaghan, with contributions from Christopher Gasson, Ankit Patel, Matthew Stiff and Jablonka Uzelac (2010) *Water Technology Markets 2010: Key Opportunities and Emerging Trends*, Global Water Intelligence.
15 US Environmental Protection Agency (2009) *Drinking Water Infrastructure Needs Survey and Assessment*.
16 www.rivernetwork.org/water-energy-nexus.
17 Ibid.

Part II

What is water tech?

Anyone who has never made a mistake has never tried anything new.
(Albert Einstein)

What do mobile phone technologies have to do with water?

How do we address the dismal global statistics on access to clean water?

It starts by thinking differently.

In Africa there are more than 500 million mobile phone subscribers[1] and, simultaneously, about 340 million people are without access to safe drinking water.[2]

How do we take a widely adopted technology such as mobile phones to address issues such as water scarcity and access to sanitation?

First, let's take a look at m.paani, a social enterprise focused on using mobile technology as a platform to provide key services – such as safe water, sanitation, education, healthcare and energy to the underserved.[3] Akanksha Hazari, a social entrepreneur, peace negotiator, and businesswoman, founded m.paani. The name of her firm is constructed from m for mobile, and *paani* is Hindu for water.

How did m.paani start and where might they be going?

In 2011, Hult International Business School teamed with the Clinton Global Initiative and Water.org to take on the challenge of providing clean water and sanitation to 100 million people in five years. The plan was for Water.org to build on their initial success in using microfinance, known as the Water Credit Program, to fund the reach and cost effectiveness of providing water and sanitation to those in need. The Water Credit Program has been successful in reducing the philanthropic cost of providing water and sanitation by about 80 percent.

The approach of bringing unconventional methods, such as microfinance, to solve difficult problems is at the heart of innovation. Water.org's approach to the 2011 competition was to identify additional trends that are "orthogonal" to the water and sanitation space – economic, technological, social, cultural and political – to increase the efficient and effective provision of water and sanitation services to the bottom of the pyramid,[4] reaching at least 100 million people in 5 years.

The key criteria for the competition were:

- Is the approach demand-based rather than supply-driven?
- Is it philanthropically efficient (philanthropic cost/person served)?
- Is it effective from the perspective of the customer (availability and accessibility)?

Other factors considered in the competition were:

- What is the cost?
- Is it consistent with local cultural norms?
- How much water is available and when?
- Can infrastructure be operated and maintained using local skills and supply chain?
- Sustainability: can the above measures be maintained over time?
- Scalability: is it based on an open-source and open-system approach?
- Is the approach actionable by Water.org in the next six months?

The winners from each of the regional competitions (Boston, San Francisco, London, Dubai, and Shanghai) met in New York City on 30 April 2011. The winner of the global competition received a $1 million donation from Hult International Business School to be used to implement the winning solution.

The global winner was the University of Cambridge from the United Kingdom.[5] Their solution was truly innovative: leveraging existing behavior by coupling their mobile phone with the existing WaterCredit platform to create a sustainable structure to provide clean water and sanitation to those at the bottom of the economic pyramid.

Akanksha Hazari was part of this team.

Now the challenge begins for m.paani and other water tech start-ups. They have secured a commitment from a major telecom to partner on a pilot in Africa and, as of mid-2012, were focusing on their first pilot project in India. Like any start-up, they will face the challenge of securing additional funding, finding partners and pilot projects, and the always daunting challenge of scaling the business to "move the needle" in addressing the challenges of increasing access to clean water and sanitation.

This is not the only mobile phone technology application to address water problems. Recently, Water for People[6] launched a project with the World Bank called Field Level Operations Watch (FLOW).[7] This open source application allows field workers to use mobile phones to document how well water pumps and sanitation points in the developing world are functioning, then transmit that data to create an online-tagged map of target regions. After successful pilot projects in Nicaragua and Rwanda, FLOW is being rolled out to more countries where Water for People operates (Malawi, Guatemala, Bolivia, El Salvador, Nicaragua, Peru, and India).

There is also the innovative mw4d at the University of Oxford in the UK, led by Dr Rob Hope and Dr Garri Clifford.[8] Mw4d is a research initiative to design and text mobile technologies to improve water security and reduce poverty in the developing world with a focus on "smart rivers, smart hand pumps and mobile payments."

And, interestingly enough, innovative thinking in addressing water challenges is not new. Humanity has been tackling the challenges of extracting, transporting and managing water for thousands of years. The challenge remains the same: how do you provide access to clean water and sanitation, as well as water for agriculture, industrial uses, domestic needs, and ecosystems, while recognizing that water is also an important part of cultural and social needs?

Notes

1   www.mobilemonday.net/reports/MobileAfrica_2011.pdf, page 5; www.wssinfo.org/fileadmin/user_upload/resources/1278061137-JMP_report_2010_en.pdf, page 7.
2   African Ministers' Council on Water (2012) *A Snapshot of Drinking Water and Sanitation in Africa. A Regional Perspective Based on New Data from the WHO/UNICEF Joint Monitoring Programme for Water Supply and Sanitation*, www.wssinfo.org/fileadmin/user_upload/resources/Africa-AMCOW-Snapshot-2012-English-Final.pdf.
3   http://mpaani.com.
4   C. K. Prahalad (2004) *The Fortune at the Bottom of the Pyramid: Eradicating Poverty Through Profits*, Upper Saddle River, NJ: Prentice Hall.
5   PR Newswire, New York, 29 April 2011, 2nd Annual Hult Global Case Challenge, http://us.mobile.reuters.com/article/companyNewsAndPR/idUS218997+29-Apr-2011+PRN20110429?irpc=932.
6   www.waterforpeople.org.
7   http://mobileactive.org/flow-where-mobile-tech-and-water-meet.
8   http://oxwater.co.uk.

Chapter 3

Do water and innovation "mix?"

Many in the water industry, entrepreneurs and investors would say no.

However, we share the contrarian view along with Sheeraz Haji, CEO of the Cleantech Group.[1] In Haji's words:

> Water and innovation don't mix. At least that's the conventional wisdom. Water's not priced appropriately. Regulation hinders the adoption of new technologies. The water market is terribly fragmented (by geography, technology, and industry); channels are difficult; investment is scarce. The people who work at water utilities and their engineering firms are conservative and don't want innovation. All of these are valid reasons why it's difficult to launch, grow, and invest in a water startup.[2]
>
> However, I believe something has changed, and water is now poised to become a top cleantech theme and investment sector over the next few years. Why?
>
> - Water pricing will (and has already) improved to more realistic levels.
> - Hidden costs (e.g. from energy costs to move water or property damage associated with water) will start to become more visible – especially to corporations.
> - Businesses will sweat the risks to reputation, product quality, up-time, and supply chain. Can a beverage factory operate without access to clean water? Can a mining company move forward with a project without the community's trust that precious water resources will be protected?
> - Corporates are investing in water. Exhibit A is Ecolab's recent acquisition of Nalco Holding or ABB's recent investment in TaKaDu.
> - Entrepreneurial talent is starting to pay attention to water, and we shall see some outstanding repeat entrepreneurs venture into the space.

However, the "water market" is not one market but in fact a series of sub-sectors. These sub-sectors are in various stages of development and each represents a unique investment opportunity. This fragmentation gives would-be entrepreneurs a good reason to pause before jumping in. However, closer examination of the sectors reveals a number of exciting opportunities (as well as challenging areas to avoid).

For example:

- Desalination has been a difficult and declining market where siting and permitting issues, high energy costs, and tough comparative economics (versus efficiency measures and reuse) have created a challenging environment for desalination start-ups.
- Demand side management represents a more attractive opportunity. The "smart water" space still lags behind the smart grid but utilities now appreciate the opportunity and are starting to invest in smart meters, software and services to bring basic analytics to their water networks. Companies like TaKaDu and WaterSmart are gaining the attention of forward-thinking utility executives. Smart Water represents a massive opportunity that's largely in front of us.
- More broadly, significant water efficiency opportunities exist across agricultural, industrial, commercial and residential sectors. There are many ways we can make better use of water across the system as well as the energy and nutrient content in our wastewaters. The biggest challenge is to figure out the business and pricing models to create value to customers who don't necessarily prioritize water as their #1 issue of the day.
- Advanced water treatment technologies are displacing conventional and chemical alternatives in many parts of the drinking water and wastewater value chains. This is driven by a broad public desire to move away from chemicals as well as industry's need for advanced water treatment. For example, the revolution in unconventional oil and gas exploration has created booming demand for products and services to treat "produced water."

While investors often complain about government's role in water markets, regulation can actually create new markets. For example, pending regulation across the globe will create a new market for "ballast water" treatment technologies and services. Ships currently move invasive species from port to port through the pumping and discharge of dirty ballast water, creating significant ecologic and economic damage across the world. Government agencies, including the International Maritime Organization (IMO), a specialized agency of the United Nations, have been developing guidelines and regulations that will dramatically alter what ship builders, owners and operators can do going forward.

While it is important to acknowledge the differences between water sectors, there are many commonalities across the water business. For example we believe there are financing challenges that cut across many water technology markets. Regardless of the substantial challenges, we are entering an exciting new investment and growth period. Cleantech Group data show investments in water rising steadily over the past eight quarters, and investor and corporate interest in water seems to be at an all-time high. The next decade should be an interesting one for innovation in water.

Again, we share this perspective: water and innovation do, in fact, "mix."

In this part of the book, we examine innovation in water supply, water data acquisition/analytics, water demand (efficiency and treatment), and the energy–water nexus. We will hear from companies, non-governmental organizations and entrepreneurs all providing their perspectives on how we are moving to new way of managing water.

The 21st century water paradigm will be about diversifying sources of water (surface water, groundwater, rainwater, brackish water and salt water) coupled with aggressive conservation, reuse and recycling. Figure 3.1 illustrates the transition from using water as a disposable commodity (left) to a valued resource (right).

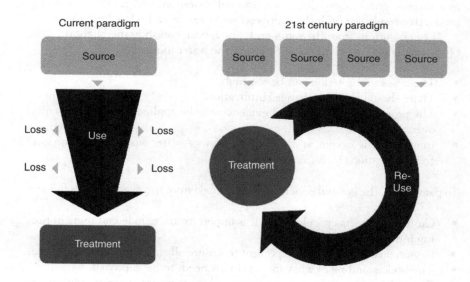

Figure 3.1 The shift to a 21st century water paradigm[3]

When one mentions water and innovation you find either optimism (such as ours and Haji's) or disbelief. Unfortunately, disbelief is fairly common, so let's examine the obstacles to increased innovation in the water industry.

Overcoming inertia

A study by Professor Martin Cave from April 2009 provides insight on the obstacles in increased innovation in the UK water markets.[4] The conclusions are informative of the state of innovation in the UK water industry and likely not unique in the water industry sector within Europe and globally. Briefly:

- Low and falling number of patents in water and wastewater technology in the UK compared to "competitors" such as Germany
- Low and falling investment by water utilities compared to other industry sectors
- Low and variable investment by water utilities

To better understand the challenges in bringing increased innovation in the water sector a conference was held in May 2012 by several stakeholder groups. "Water Innovation Europe" was the joint stakeholder event of the Water supply and sanitation Technology Platform (WssTP) and the EUREKA Cluster for water (ACQUEAU), held in Brussels in May 2012.[5] The conference was viewed as a success, encouraging discussion and collaboration, and resulting in highly productive results for both the European water sector and related industries.

The conclusions from the water and innovation workshop shed light on some of the challenges in driving innovation in the water industry:

- The cost of innovation can be too high.
- There should be "step change" innovation.
- The process of persuasion and evidence can be applied to drive innovation over time.
- Innovation is active at the intersection of water and information and communications technology (ICT).

To paraphrase, the general conclusions developed from the event were as follows:

- Customer aspects are relevant – it is important to include end-users in testing innovative technologies.
- Regional collaboration is important to ensure all stakeholders are included.
- The understanding of what innovation is needs to be improved.
- There should be a creation of a protected space where experimentation can take place and technologies/approaches can be tested – allow for learning from failure and share risk of failure.
- It is important to re-engage people with water and create a new water culture.

Hugh Goldsmith, head of water and waste management at the European Investment Bank (EIB), captured the key barriers well in his presentation, "EU

Water Sector Financing and Innovation."[6] The top three barriers to investment in research and development (R&D) were: excessive risk, excessively high innovation costs, and lack of sources of finance. Other factors identified are: lack of qualified staff, lack of information and regulatory issues.

The "crisis" of innovation in water and wastewater is not a new idea – perhaps dating back thousands of years. A more recent perspective on the barriers for adoption of innovation in the water industry was appropriately titled *The Crisis Of Innovation In Water And Wastewater*.[7] The thesis of the authors was that while innovation was alive and well during the twentieth century, innovation in the water industry was stuck in the Victorian era – "pouring concrete and harnessing outmoded technologies and management approaches."

The authors argue that when there has been multi-stakeholder "setting of standards and technical parameters, agreement of needs and priorities, and tailoring of the 'pitching' or 'packaging' of an innovation to the risk-averse culture of the sector to enhance its appeal to various professional/disciplinary constituencies", events "appear to have gone well." The approach of an "open and informed partnership during demonstration and scale-up of promising inventions then seem to have often gone on to become accepted, successful innovations in wide-scale use."

The authors go on to clarify that it is "innovation management" and related R&D that is successful. Not just technology innovation, but a new approach to stakeholder dialog and solution development (essentially, the previously discussed "collective action").

The authors conclude that:

- "Innovation is the exception, not the norm: saying 'the water sector has been very innovative' when one means 'the sector water has been very busy with business as usual' (or has just spent a lot of money) is unhelpful."[8]
- "Innovation can fail – with adverse consequences for an individual, group, organisation, and whole economic sector or beyond. It will fail far more often than it will succeed. Some people also see it as an amoral process. That may mean it does not have socially or environmentally beneficial outcomes when it succeeds. People focus disproportionately on innovation success."[9]
- "Innovation is not just about technical devices, etc.; it's about products, processes, services or knowledge new to the world or to a sector – this includes management and cultural practices by definition, so there's no need to talk of them separately and assume people mean gadgets and widgets when they mention 'innovation'."[10]

Clearly, what has been lacking historically in adopting innovation in the water industry has less to do with actual technologies but, instead, developing innovative stakeholder engagement processes to build consensus on the need for and approaches to a more efficient and effective use of water.

Fortunately, there are hopeful signs.

Global initiatives

When it comes to innovation (technology and partnerships), innovation is coming from an increasing number of organizations, innovation hubs and countries committed to addressing their local needs and exporting technologies as a business opportunity. These countries and organizations are focused on raising the awareness of water issues and driving innovation, and are having an impact on developing technology and policy solutions.

While just a sampling, the following will give you a feel for the organizations and countries pushing the water technology innovation agenda.

We will start with Israel and Singapore, widely recognized as global leaders in addressing their own water-related challenges and building a water technology export industry.

Israel

One could make the case that Israel is a global leader in water tech innovation out of sheer necessity and "entrepreneurial DNA." The country has an entrepreneurial culture; a strong scientific community and water scarcity focuses the need to drive water technology innovation.

According to the "Invest in Israel" program produced by the Israeli government, since its foundation in 1948, Israel has placed great emphasis on maximizing its water supply, famously turning much of its arid land into fertile agricultural soil.

David Ben Gurion, Israel's founding father, declared the goal of "making the desert bloom" as one of the central themes of the new nation, believing it could be one of its main contributions to the world.

Thus, water technologies have been a national priority in Israel since day one. This emphasis has proven itself in one of the world's most efficient and innovative water systems. To that end, the Israeli water industry is today recognized as a global leader in the water technology arena thanks to breakthrough innovations in areas such as desalination, drip irrigation and water security.

The government supports water investments through a program titled NewTech:

> Israeli NewTech was founded on the belief that the Israeli water and renewable energy sectors have the talent and capability to be strong growth industries for the country, and to play an important part in establishing the "Next Generation Oasis" for the world's rising needs. This pioneering national program is led by the Ministry of Industry, Trade and Labor, and is supported by a number of additional Israeli government agencies. Israel NewTech helps to advance the water and renewable energy sectors by supporting academia and research, encouraging implementation in the local market, and by helping Israeli companies succeed in the international arena.[11]

According to an article by Genevieve Long titled "Israel's Water Innovation Leading the World", Booky Oren, the chairman of the Israeli company Miya, viewed water as "a big business" and, as a result, translated his 25 years of experience into growing the water cleantech sector in Israel. Mr Oren was formerly the chairman of the board of Mekorot (Israel's national water company) until 2006. His perspective captures the essence of the business opportunity – in 80 years, Israel's population grew 25 times, but "there's no way the water grew by 25 times."[12]

The water tech industry in Israel grew from necessity. Israel recycles 70 percent of its wastewater, which is three times the quantity of its nearest competitor, Spain. Desalination plants supplying 500 million m3 of water per year will be providing about 35 percent of the freshwater needs of the country by 2013.[13] Mr Oren believes that the process of taking seawater and making it potable for human consumption is the wave of the future.

Mekorot not only is responsible for the operational aspects of ensuring the country has adequate supplies of water, but also is focused on finding better, faster and new methods to sustainably manage water resources. This means that on occasion they will establish international relationships to share technology and skills. Mekorot has initiated a three-year relationship with the Argentinean company 5 De Septiembre to share water delivery and treatment technologies.

There is no shortage of stories of how Israel is leading in water tech innovation. A May 2011 article by *The Economist* accurately characterized water tech innovation in Israel: "Israel wants to become the Silicon Valley of water technology."[14]

A few statistics highlight their success:

- 2008 output equaled US$1.4 billion.
- 2011 output was projected to double to $2.5 billion.
- Every $1 invested in sanitation or drinking water returns $3 to $4 in economic returns.
- The global water market, worth $500 billion annually, is growing at 7 to 8 percent.[15]

By these numbers, in 2011, Israel captured about 0.5 percent of the market.

Israel is the home of leading companies in drip-irrigation technologies such as Netafim. Netafim leads the smart irrigation technology revolution not only in Israel, but exports their technology. This is part of Israel's "National Program for Promoting Water Technologies," launched in 2006, which promotes the industry internationally by investing in R&D, deploying technology in local and global marketplaces, and linking domestic companies with foreign companies.

In 2006 the Israeli government launched a program to support water companies, for instance by helping them market their products abroad. It also created (and later privatized) Kinrot Ventures, the world's only startup incubator specializing in water technologies. In November 2012, Hutchison Water (Israel) Ltd won the tender to operate Kinrot Technology Ventures for the next eight years. Hutchison Water Israel said it will invest 100 million shekels in Kinrot's portfolio companies over the next eight years. According to Hutchison's representative in Israel, Dan Eldar, "The acquisition of Kinrot will give new, promising Israeli companies access to the highest levels of expertise in the water and cleantech industries, and will help them reach a range of foreign markets where Hutchison operates."[16]

As an indication of the water industry activity in Israel, Hutchison Water's acquisition of Kinrot is part of the Office of the Chief Scientist tender for the five new franchisees that will operate incubators for the next eight years.[17]

There have been a number of Israeli water tech start-ups including the following.

- Netafim got its start on a kibbutz in the Negev desert. Intrigued by an unusually large tree, an agronomist discovered that a cracked pipe fed droplets directly to its roots. After much experimentation the firm was founded in 1965 to sell what has become known as drip irrigation.
- Aqwise provides equipment and expertise to build wastewater treatment plants. Facilities based on the firm's technologies feature what it calls "biomass carriers", thimble-sized plastic structures with a large surface area. In wastewater pools they give bacteria more space to grow and thus allow biological contaminants to be consumed more quickly.
- Emefcy is focused on reducing the energy required to treat contaminated water, which currently uses up to 2 percent of the world's power-generating capacity. One of its products uses "electrogenic" bacteria to turn wastewater pools into "batteries."
- TaKaDu targets leaks in a water-supply network, sometimes before they happen. It does this by data analytics from the network's sensors to look for anomalies. A persistent 1 percent flow rate can identify a leak. TaKaDu's detection data analytics are monitoring water-supply systems in a dozen places, including London and Jerusalem.

Israeli companies are also exporting technologies to countries such as Argentina and India. For example, IDE Technologies has set up 22 thermal desalination facilities in India for use in nuclear power generation, oil refineries and industrial plants. There are numerous examples: Amiad Filtration Systems is working through its subsidiary Amiad India to offer eco-friendly filtration solutions for industrial, municipal and agriculture applications; AYALA Water & Ecology is using its phytotechnology concept, called Natural Biological System (NBS), to treat water, soil and air pollution; and Netafim's subsidiary Netafim India has two manufacturing plants in Vadodara and Chennai.

Water tech is big business in Israel, and getting bigger by design.

Singapore

Singapore has emerged as a leader in developing innovative technologies in the water industry and establishing itself as a catalyst for collaboration in water innovation.[18] The 45-year journey by Singapore in sustainable development and becoming a global leader on water practices and technologies is chronicled in *The Singapore Water Story*.[19] Singapore's relentless drive for water self-sufficiency in an urban context continues to shape their development policies and agendas through the coming years.

Singapore made a very deliberate effort to address water scarcity and become a leader in addressing key water industry issues. The government created a comprehensive environmental management system, including water supply, control of river pollution, establishment of well-planned industrial areas, and a world-class urban sanitation system as the starting point. This became the foundation for their goal of a "sustainable water supply" for the island country. The country's move towards a sustainable water supply consists of several key initiatives – all of which can be adopted by countries and governments.

- *Political commitment*. The Singapore government is committed to implementing a long-term water strategy and the Prime Minister supports their Four Taps Strategy (see below).
- *Institutional integration*. The Ministry of Environment and Water Resources has full responsibility for water-related affairs, including policy formulation, planning and infrastructure, and eliminates administrative barriers in water management making implementation effective and efficient.
- *Integrated land use planning*. Singapore is effective in integrating land use planning and water management – the two are inseparable. The benefits include preventing water pollution and effective management of catchments. Moreover, Singapore is successful in coordinating across relevant government agencies in water management. This ensures success of integration, and reduces inter-sectoral conflicts of interest.
- *Enforcement of legislation*. Strict implementation of legislation such as pollution control is another essential characteristic of water management in Singapore.
- *Public education*. School education and public campaigns are used to increase public awareness of water policy and programs. These activities boost public support for the government's water policy and initiatives.
- *Water tech innovation*. Singapore invests, deploys and exports innovative water technologies.

These key initiatives are driven by Singapore's strategy to become self-reliant on water, and this has resulted in a rethinking of the most basic aspects of water: sources of supply, treatment, reuse, public policy and pricing. A cornerstone to Singapore's approach to water stewardship is what the Public Utilities Board

(PUB) terms its "Four Taps Strategy." The four taps are: water reclamation, desalination, water efficiency and importation. This strategy also captures current thinking on source diversification; catchment harvesting, importing of water from Malaysia, water reclamation and desalination.

The two new "taps" are water reclamation and desalination. Reclaimed water is marketed as "NEWater." Using membrane ultra-filtration, reverse osmosis and UV treatment, this recycled water is successfully integrated into the national water supply, initially for non-potable uses, and then blended with reservoir water for potable purposes. Singapore has a total of five PUB-built plants that meet 30 percent of the state's water demand. By 2060 NEWater is projected to meet 50 percent of water demand. Singapore has taken recycled water a step further, selling bottled NEWater to increase consumer confidence in recycled water.

Singapore is also investing in desalination, its fourth "tap." Opened in 2005, the SingSpring SWRO plant is Singapore's first desalination facility, and one of the most energy efficient in the world. Water was priced at $0.48 per m^3, which was a record low for desalinated seawater when the plant opened. This plant has the capacity to produce approximately 136 million liters a day, which represents approximately 10 percent of the national water requirement.

In 2007, PUB won the Stockholm Industry Water Award as an example of integrated water management encompassing not only sound public policy but also innovative engineering solutions. This award reflects how well Singapore is moving toward its goal of being water self-sufficient by 2061.

For Singapore, water is not just a strategic risk; it is also a business opportunity. The country is promoting water-technology development: Nanyang Technological University has three water-related research units, and Singapore's water industry now has more than 50 companies. These companies are winning international contracts based on their water know-how.

Most importantly, the country wants to be a hub for water innovation and has established a center for water excellence known as "WaterHub." The WaterHub was launched in 2004 and is focused on technology, learning, and networking to build a sustainable water industry in Singapore.[20] According to the World Economic Forum, by 2015 Singapore's water industry is expected to contribute US$1.2 billion to the country's GDP, and 11,000 new jobs creating a global center of water industry expertise.[21]

The Ministry for the Environment and Water Resources, the Singapore Water Association (SWA), and Water Network launched the WaterHub. The WaterHub focuses on technology, learning and networking and is intended to be a strategic platform for the Singapore PUB and the national water industry. A key alliance of the WaterHub is also the Environment & Water Program Office (EWI) which was established in 2006 by Singapore's Ministry of the Environment and Water Resources.

The goal of the WaterHub and Singapore's PUB Technology and Water Quality Office (TWQO) is to become a water research and development incubator center for the emerging water industry. This water innovation ecosystem in Singapore is seeking to increase water resources, keep water costs competitive, as well as manage water quality and security for the Singapore government. More than 70 water companies and 14 corporate research and development centers have already set up facilities and offices in Singapore.

Singapore has also positioned itself as a hub for thought leadership on water innovation. The annual Singapore International Water Week (SIWW) increasingly draws large international crowds focused on addressing complex water issues. The 2012 event was no exception with the theme of "Water Solutions for Livable and Sustainable Cities." According to SIWW, the 2012 event "achieved a new record of S$13.6 billion in total value for the announcements on projects awarded, tenders, investments and R&D MOUs made at the event. The event drew 18,554 participants from 104 countries / regions, and about 750 participating companies."[22]

The 2012 event highlighted collaboration on water technology innovations, including:

- Singapore national water agency PUB's memorandum of understanding (MOU) with MEIDEN.
- The first Ceramic Membrane Bioreactor (MBR) Demonstration Plant to recycle industrial used water.
- Signing of a Research Collaboration Agreement formalizing the alliance between DHI, NTU, and Suez Environnement Office to make used water treatment energy-neutral.
- PUB's collaboration with Mitsubishi Heavy Industries Ltd (MHI) on the research and development of integrated water infrastructure system for used water treatment and reclamation.
- memsys's MOU with Senoko Energy for the construction of a membrane distillation system based on its novel vacuum multi-effect membrane-distillation process.
- Saudi Arabia's National Water Company announced that they would be investing approximately S$11 billion on capital expenditure on municipal water infrastructure such as water and used water treatment plants, networks and mains, for four major cities – Riyadh, Jeddah, Makkah and Taif – over the next five years.
- The Metropolitan Waterworks and Sewerage System (MWSS), the Philippines' leading water authority, announced its US$1.5 billion investment program to establish a Water Security Legacy (WSL) for the 15 million residents of Metro Manila while Singapore-based United Engineers Limited (UEL) announced three environmental engineering contracts worth over S$70 million, including two projects at the Changi Water Reclamation Plant and one for a waste-to-energy project at a poultry farm in Singapore.
- Singapore water company Hyflux also officially launched its Hyflux Innovation Centre.

Singapore is also expanding its reach, including collaboration with the International Water Summit in Abu Dhabi, the Dutch water industry, and the GWI to organize discussion platforms for specific topics.[23]

The competition to lead in water tech is increasingly formidable. Singapore also wants to be a water tech hub, and water tech clusters are emerging in other countries around the world.

Australia

Australia has led in developing water tech innovation out of necessity. Not just in technology but in a restructuring of their water industry – essentially innovation in public policy and pricing. The National Water Initiative (NWI) is Australia's enduring blueprint for water reform. The Intergovernmental Agreement on a National Water Initiative was signed in June 2004. The NWI is a shared commitment by Australia's governments to "increase the efficiency of Australia's water use, leading to greater certainty for investment and productivity, for rural and urban communities, and for the environment."

Under the NWI, governments have made commitments to:

- "prepare water plans with provision for the environment;
- deal with over-allocated or stressed water systems;
- introduce registers of water rights and standards for water accounting;
- expand the trade in water;
- improve pricing for water storage and delivery; and
- meet and manage urban water demands."[24]

The NWI's overall objective is to achieve a nationally compatible market, regulatory and planning based system of managing surface and groundwater resources for rural and urban use that optimizes economic, social and environmental outcomes. The NWI includes objectives, expected outcomes and commitments for eight inter-related elements of water management. And, each state and territory government is required to prepare an NWI implementation plan.[25]

This national policy directive on water management has set the stage for continuous water tech innovation from desalination, reuse, recycling, advance metering and efficiency.

European Union

The European Union (EU) has committed to addressing sustainable water programs for the past several years. The overall objective of EU water policy is to ensure "access to good quality water in sufficient quantity for all Europeans, and to ensure the good status of all water bodies across Europe." While Europe is considered as having adequate water resources, water scarcity and drought are increasingly frequent and widespread in the EU. According to the EU, further deterioration of the water availability is expected if temperatures keep rising as a result of climate change.[26]

As with other countries and regions experiencing the impact of water scarcity, EU countries are driving water tech innovation. For example, Denmark is home to about 200 water technology companies, employing about 35,000 people. The Danish water sector has a significant potential for increased technology exports to the BRIC countries (Brazil, Russia, India and China), along with countries such as Bangladesh, Egypt, Indonesia, Iran, South Korea, Mexico, Nigeria, Pakistan, the Philippines, Turkey, and Vietnam. The Danish Foreign Ministry has appointed export ambassadors to support the export potential for Danish industries for some of these growing markets and has secured demonstration projects in India and China. These projects include a demonstration project for sewage treatment in India, a project on improved sludge management in India as well as the "Innovative Green System Solutions" project focused on the reduction of non-revenue water in Indonesia.[27]

Denmark is just one example of how EU governments are promoting the development and export of water tech innovation.

United States

The USEPA is increasingly committed to promoting innovation in the water industry. The Water Technology Innovation Cluster (WTIC) formed in January 2011 in the Dayton/Cincinnati/northern Kentucky/southeast Indiana region of the US to develop and commercialize innovative water technologies that solve environmental challenges and promote sustainable economic development in the region: "The WTIC builds on existing firms, intellectual capacity, and expertise in the region that can be used to advance economic development and technology innovation in a strategic and coordinated manner."[28]

The regional technology cluster is a geographic concentration of firms and supporting institutions that are committed to building a vibrant, technology-driven economy with a particular focus area: water. The region was selected because it contains several of the elements needed for a successful cluster, including USEPA's water research laboratory, and progressive water utilities such as the Greater Cincinnati Water Works (GCWW). The WTIC's initial focus will be on drinking water, with $5 million in USEPA Science to Achieve Results (STAR) grants dedicated to clean drinking water and the formation of the National Center for Innovative Drinking Water Treatment Technology.

The WTIC has formed work groups to advance the WTIC. Representatives of the EPA cluster team are participating in six of these work groups. Participation can include providing technical input on specific issues, serving as a conduit between EPA Cincinnati and WTIC, and identifying joint activities to pursue with WTIC. The work groups targeted the following activities.

- *State testing protocol* – To facilitate the development of a protocols for test and approval of water devices in partnership with organizations across the three-state cluster region.
- *Network of test beds* – To coordinate the development of and access to a network of test bed sites to evaluate local, national, and international water technologies.
- *Water policy forum* – To convene a water policy forum that will identify issues or questions within the water industry, facilitate dialog with USEPA or other regulatory agencies on those questions, and report key knowledge back to the water community.
- *Seminars, conferences, workshops* – To develop a premier series of water-related educational programming, including technology workshops, industry or technology seminars, and water conferences.
- *Business advisory council* – To establish a business advisory council of businesses and inventors/entrepreneurs with water-related products.
- *Top three to five technology problems identification* – To define the top three to five technology problems to be addressed to better serve the water community and to improve the competitive advantage of businesses within the WTIC's region.
- *Partnerships with other technology/research entities* – To use the WTIC as a hub for developing partnerships with other technology, cluster, and research organizations.
- *Establish success/impacts metrics* – To establish the metrics that will be used to measure the success and impact of WTIC overall.
- *Communications and marketing* – To develop a top-notch communications and marketing plan capable of communicating WTIC impacts worldwide.
- *Coordinate joint activities* – The USEPA cluster team coordinates joint activities (such as meetings, workshops, and test events) that support EPA and the WTIC's interest in R&D for innovative water technologies that advance economic development and solve environmental challenges. As part of this role, the cluster team coordinates both internally (with EPA Cincinnati research staff) and externally (with the WTIC) to identify topics, and capabilities.[29]

Canada

There is seemingly no shortage of water in Canada, yet they have recognized that they have expertise in water tech and have a committed strategy to export this technology. Water and water technology "know-how" is viewed in Canada as an asset. Insight as to how Canada views water and water tech is outlined in *Water and the Future of the Canadian Economy*, a summary document that captures the thoughts from the Canadian Water Summit held on 17 June 2010.[30]

The report has been a driver in understanding the intersection of water and the Canadian economy. The goals of the report were to:

- Build understanding around the central role water plays throughout the Canadian economy, with a view to improving how management of this vital resource can enhance national productivity and competitiveness;
- Bridge the gap between research, policy and management practices to help diverse economic actors find a common platform for collaboration; and,
- Highlight strategic areas where creative, proactive intervention can position Canada as a progressive and influential voice for addressing water issues within and outside our borders.[31]

Let's focus on the last objective – water tech innovation as an export. The report states that:

> the global water industry presents significant opportunities for Canadian companies and that Canadian water sector firms maintain a technical lead in such areas as purification, membranes and hydrogeology, and have substantial expertise in traditional water services like consulting, engineering, quality analysis and construction.[32]

The focus of exports should be the US, as well as fast-growing economies in Asia and Latin America to address infrastructure needs.

The report identifies several challenges that must be overcome and recommendations – "greater attention is required among policymakers, businesses and investors to ensure Canadian enterprises can participate meaningfully in the global marketplace."[33]

The Canadian Province of Ontario was listening. The Ontario Ministry of Economic Development and Innovation has established the Water Technologies Acceleration Project (WaterTAP) seeking to build upon the local academic and commercial enterprises focused on the water industry. The WaterTAP was established under the Water Opportunities and Water Conservation Act, 2010, to help expand globally competitive companies, and provide high-value jobs in Ontario's water and wastewater sector. WaterTAP will identify opportunities, build partnerships, and improve information exchange within the Ontario water and wastewater sector.[34]

WaterTAP will also focus on attracting new, international investors and potential customers to Ontario to gain access to leading research and the world's brightest water experts. WaterTAP achieved success in February 2011 when Singapore and the Government of Ontario agreed to share technology and expand their respective water industries. Singapore's national water agency, PUB, and the Government of Ontario, Canada, agreed to enter into a strategic alliance to conduct advanced clean water research and development.[35]

Mexico

Water scarcity and water quality issues are increasingly being felt in Mexico. Water resources management is one of Mexico's pressing concerns, and it is imposing heavy costs to the economy. The arid northwest and central regions contain 77 percent of Mexico's population and generate 87 percent of the gross domestic product (GDP). Southern regions have abundant water resources. Surface and groundwater are overexploited and polluted, leading to insufficient water availability to support economic development and environmental sustainability. The country has put in place a system of water resources management that includes both central (federal) and decentralized (basin and local) institutions.[36]

The Mexican government recognizes these challenges and is adopting innovative solutions. One such example is the recent announcement that The Water Initiative (TWI) technology solutions will be allowed to deploy point-of-drinking (POD) or point-of-use (POU) water systems to meet local needs. TWI's approach is to:

> engage local communities to customize and create comprehensive and sustainable technology solutions which effectively remove water contaminants such as pathogens (bacteria and viruses), unsafe levels of inorganic materials (such as arsenic and fluorides) and other harmful chemicals and contaminants of concern.[37]

In April 2012, TWI launched its POD filtration solution, WaterCura®, in the State of Durango, Mexico. TWI describes this as "a key first step in the process of showing that centralized water treatment is not the only answer in providing contaminant free and affordable clean drinking water at the point of where it is consumed or used."[38]

This technology deployment follows the November 2011 decision by the State Water Commission of Durango, which awarded TWI initial contracts to manufacture and install over 30,000 innovative POD devices for rural and urban homes to address water contamination issues in the area.

According to PRweb, TWI's launch in Mexico was attended by Durango's governor Jorge Herrera Caldera, CONAGUA director general of North Central Region Cecil Castro, director general of water for State of Durango Miguel Calderón Arámbula, director of SIDEAPA José Miguel Campillo Carrete, and the mayor of Gómez Palacio Rocío Rebollo Mendoza.[39] TWI's solution was 30 percent of the cost of centralized treatment and 16 percent of other POD proposals submitted in response to the public bidding process. In total, over 50 rural communities and 50 urban communities across four of Durango's municipalities (equating to a population of nearly 150,000) will benefit from safe, clean drinking water through the WaterCura installations. The government of Mexico has certified TWI's products as the appropriate solution for their water challenges. In fact, earlier work in Mexico by TWI resulted in being awarded "The Global Game Changer" in 2011 by the EastWest Institute (EWI) for its work in developing public–private partnership solutions to address the drinking water contamination issues in Mexico.

TWI's work is truly innovative. Watch how TWI gains traction in Mexico and elsewhere globally.

China

China is thirsty for water. As we discussed, the key drivers in water scarcity are increased population, urbanization, the rise of the middle class and economic development. China is at the nexus of all of these issues.

According to an article by Kacj Perkowski, "China's water supply is smaller than that of the US, yet it must meet the needs of a population nearly five times as large. Industrialization has taken its toll on this already limited resource."[40]

A few key points from the article:

- "Industrial and biological pollution has contaminated almost 90 percent of the underground water in Chinese cities.
- The World Health Organization (WHO) estimates that one in four (300 million) Chinese do not have daily access to clean water, and that roughly half (700 million) are forced to consume water below WHO standards.
- High population density, a poor ratio of available water to demand, and regional imbalances in available water supplies are serious challenges for China in managing its usable water supply.

China is now aware that water scarcity and pollution need to be addressed. Their recent five-year plan will shift its environmental focus to water:

> From 2011 to 2015, the country will spend a total of USD$536 billion on water purification and waste water treatment plants, irrigation systems, and flood control projects. Currently, only 50 percent of urban sewage is treated. By 2015, the government intends to add 42 million tons of daily sewage treatment capacity to increase its urban waste water treatment rate to 85 percent.[41]

All of this means investment in water tech innovation. A few examples of the drivers and opportunities from the Perkowski article are provided below:

- Over the next five years, it is estimated that an additional investment of RMB 220 billion ($35 billion) will be needed to upgrade existing purification facilities and bring them into compliance.
- Water resources utilization will limit the annual consumption of water to 635 billion cubic meters by 2015, further increasing the need for water recycling facilities. China expects to spend $69 billion on industrial wastewater treatment. Because per capita water resources in China are only a quarter of the world's average, and industrial water consumption constitutes a quarter of the country's total water consumption, the recycling of deeply treated industrial waste water is essential.

- Implementation of the water-related programs called for in the 12th five-year plan has already begun. In 2011, the first year of the plan, total spending on water resources management increased significantly to RMB 345.2 billion ($54.6 billion). In addition to water treatment and recycling, China has already initiated programs to limit the loss of human life and property damage caused by flash floods. At the end of 2011, RMB 3.8 billion ($603 million) was earmarked to subsidize flash-flood forecasting projects in 1,100 counties throughout the country. It is expected that the number of counties will be increased to 1,800 and that $1.8 billion will be spent on flash-flood forecasting programs by 2013.[42]

And to bring the discussion full circle, Israel is selling water technology to China.[43] Israel has sold to China water technology valued at $300 million for use in the agriculture sector. There will be more to follow, as China scrambles to meet their increasing water needs and clean up what resources they have.

Other initiatives

Shifting gears just a bit, it is also important to note the work of the Stockholm International Water Institute and Circle of Blue in increasing awareness of the need to address water issues and bringing together stakeholders to develop innovative solutions to these challenges. While not country-led initiatives, they are increasing awareness of water-related risks and opportunities.

Stockholm International Water Institute

The Stockholm International Water Institute (SIWI)[44] is a policy institute that seeks sustainable solutions to the world's water issues. SIWI manages projects, synthesizes research, and publishes findings and recommendations on current and future water, environment, governance and human development issues.

From the SIWI website is the following description:

> SIWI serves as a platform for knowledge sharing and networking between the scientific, business, policy and civil society communities. SIWI builds professional capacity and understanding of the links between water–society–environment–economy.[45]

SIWI also hosts the annual World Water Week, which brings together a diverse group of stakeholders in the water industry, and awards the annual prestigious Stockholm Water Prize. The importance of the Stockholm Water Prize is captured by the 2012 winner, PepsiCo's Dan Bena:

> On June 13, 2012, PepsiCo was named the recipient of the prestigious Stockholm International Water Institute's 2012 Stockholm Industry Water

Award. The award recognizes the company's commitment to reducing water consumption in its operations and efforts to help solve water challenges on a broad scale. The company's efforts to increase water efficiency were recognized by the Stockholm Industry Water Award jury. According to the jury, PepsiCo has set and achieved a high standard for its own operations, and has demonstrated that responsible water use makes good business sense.

Agricultural water conservation is a critical element in the water/food security nexus, and our stewardship efforts are crucial in securing a resilient supply chain for our business and helping the communities where we operate to thrive. We are proud to work with a variety of exceptional partners that help us reach aggressive water stewardship goals.

PepsiCo is incredibly proud to win the prestigious Stockholm Industry Water Award since the Institute has developed a world-class standing for credibility, innovation, and thought leadership in the water space. They understand the importance of engaging the private sector along with government, NGOs, academia, and other partners to work toward common and collaborative solutions to the global water issues.[46]

Circle of Blue[47] is an effort by the Pacific Institute[48] to bring the world of journalism to the world water crisis.[49] Increasing awareness of the risks and opportunities from increasing water scarcity is the first step in driving innovation in the water industry. While not focused on technology innovation per se, Circle of Blue is a voice for the private and public sector challenges of water scarcity, water quality and access to clean water and sanitation. In the ever-increasing information overload, Circle of Blue provides a timely summary of critical water news.

These countries, water technology innovation clusters, events and Circle of Blue are transforming the water industry. The clusters work on fostering innovation in the broad areas of increasing water supply, efficiency, water treatment, water infrastructure repair, and in data acquisition/analytics.

Water as part of a bigger issue – resource scarcity

While we are focused on innovation in the water sector, it is important to note that water is part of a broader challenge in resource innovation, which is essentially how water is tied to energy, land use, and food. This integrated view of resource innovation/productivity is the real opportunity for the private and public sectors as the world faces increasing population growth and demands on key resources.

Read or listen to the Circle of Blue interview with Parag Khanna titled "How Resource Scarcity Will Lead to a New Global Order."[50] Now a few years old, the perspective is nevertheless critical in understanding how resource scarcity (including water) will drive innovation.

"Business as usual" is not guaranteed.

A recent report from Europe also illustrates this point. According to a report by SciDev, which is funded by the European Commission and seven European states:

> investment in innovation is required for sustainable agriculture, for achieving more efficient use of water and energy, and for rolling out renewable energy technologies. More technological innovation is needed to fight growing resource scarcity, but it will only be successful in achieving sustainable development if it considers the use of water, energy and land as interdependent issues.[51]

A few key points and quotes from the report are worth noting to drive home the point that an integrated resource approach is needed to tackle the increasing global demands on water, energy and food:

- Failure to consider the three basic resources of water, energy, and land as a "nexus" – in which the use of one affects the availability of the other two – is leading to poor decisions that ultimately work against sustainable development.
- The interconnectedness of resources is a phenomenon the authors believe is largely ignored in setting policy.
- There is an inability to solve resource issues by tackling one problem at a time.

The report's authors also call for institutions that govern access to land, water or energy at local, national and regional levels to be established ("or radically reformed"), to take into account the water–energy–land nexus. One of the goals is to push the conversation on water innovation to one that recognizes that integrated resource innovation is desperately needed now (more on this topic in Chapter 5 on the energy–water nexus).

Specifically, what might innovation in the water sector look like to address the broader issue of resource scarcity? A few key themes to watch in water innovation are:

- smart water (monitoring, sub-metering, precision irrigation, network optimization, plant and facility management);
- on-site treatment and use; and
- extracting energy and resources from wastewater sludge.

From the perspective of CitiGroup, the top trends to watch are:

- desalination;
- water reuse;
- produced water;
- membranes displacing chemicals;

- ultraviolet light disinfection;
- forward osmosis used in desalination;
- water efficiency products; and
- point-of-use treatment.[52]

So let's break the water innovation discussion into a couple of basic parts and examine the drivers and opportunities within each subsystem. The basic parts are water supply and water demand, with the demand discussion focused on efficiency and treatment (reuse and recycling).

Notes

1   www.cleantechgroup.com and Sheeraz Haji, personal correspondence for this book.
2   Ibid.
3   Adapted from Lux Research Water Intelligence (2008) *Water: Evolution & Outlook of the Hydrocosym*, www.e-zcoursebooks.com/WaterEnergyNexus.pdf (accessed 4 December 2012); and from William Sarni (2011) *Corporate Water Strategies*, London: Earthscan.
4   Martin Cave (2009) *Independent Review of Competition and Innovation in Water Markets: Final Report*, Warwick Business School, April.
5   www.waterintheurbanenvironment.eu/?page_id=37.
6   Ibid.
7   Duncan A. Thomas and Roger R. Ford (2005) *The Crisis Of Innovation In Water And Wastewater*, University of Salford.
8   Ibid.
9   Ibid.; but here the authors note that "but happily there's a new book on failure cases, *Challenging the Innovation Paradigm* (we didn't go too 'pro-innovation' in our UKWIR work, thankfully, as the research design involved success/failure pairs to tease out why differing developments that could address the same issue made it or were abandoned)." The book they mention is edited by Karl-Erik Sveiby, Pernilla Gripenberg, and Beata Segercrantz, and published by Routledge in 2012.
10 Ibid.
11 http://israelnewtech.gov.il/English/Pages/default.aspx.
12 Genevieve Long (2010) "Israel's Water Innovation Leading the World", *The Epoch Times*, 7 August, www.theepochtimes.com/n2/world/israels-water-innovation-leading-the-world-40539.html.
13 www.financialexpress.com/news/israel-rides-high-on-water-market/793963/0.
14 www.economist.com/node/18682280?story_id=18682280.
15 www.financialexpress.com/news/israel-rides-high-on-water-market/793963/0.
16 www.israelnewtech.com/2012/11/kinrot-hutchison.
17 www.globes-online.com, 11 November 2012.
18 http://siteresources.worldbank.org/INTEAPREGTOPENVIRONMENT/Resources/WRM_Singapore_experience_EN.pdf.
19 Cecilia Tortajada, Yugal Kishire Joshi and Asit K. Biswas (2013) *The Singapore Water Story: Sustainable Development in an Urban City State*, Abingdon: Routledge.
20 www.pub.gov.sg/waterhub/Pages/default.aspx.
21 The World Economic Forum Water Initiative (2011) *Water Security: The Water–Food–Energy–Climate Nexus*, Washington, DC: Island Press, page 215.
22 www.siww.com.sg.
23 www.greenbusinesstimes.com/2012/07/07/singapore-international-water-week-2012-closes-on-a-high-press-releases.

24 www.environment.gov.au/water/australia/nwi.
25 www.environment.gov.au/water/australia/nwi/index.html.
26 http://ec.europa.eu/environment/water/quantity/scarcity_en.htm.
27 www.danishwaterforum.dk.
28 http://watercluster.org/wordpress.
29 Ibid.
30 *Water and the Future of the Canadian Economy*, www.blue-economy.ca/sites/default/files/reports/resource/CDN_Water_Report_PDF_English.pdf
31 Ibid.
32 Ibid.
33 Ibid.
34 www.mri.gov.on.ca/english/programs/waterTAP.asp.
35 www.mri.gov.on.ca/english/news/WATER021711.asp.
36 http://en.wikipedia.org/wiki/Water_resources_management_in_Mexico.
37 www.thewaterinitiative.com.
38 www.prweb.com/releases/water-contamination/clean-water/prweb9440241.htm.
39 Ibid.
40 Kacj Perkowski (2012) "Quenching China's Thirst For Water," *Forbes*, 23 April, www.forbes.com/sites/jackperkowski/2012/04/23/quenching-chinas-thirst-for-water.
41 Ibid.
42 Ibid.
43 Reuters, 29 February 2012.
44 www.siwi.org.
45 Ibid.
46 Dan Bena, personal correspondence for this book.
47 www.circleofblue.org.
48 www.pacinst.org.
49 www.circleofblue.org/waternews.
50 www.circleofblue.org/waternews/2010/world/parag-khanna-how-resource-scarcity-will-lead-to-a-new-global-order.
51 SciDev (2012) *European Report on Development 2011–2012*, 25 May, www.scidev.net/en/science-and-innovation-policy/science-at-rio-20/news/innovation-must-consider-water-energy-and-land-jointly-.html.
52 CitiGroup (2011) *Citi Water Sector Handbook*, May 24.

Chapter 4

Water supply

Historically, meeting water supply demands was met by increasing the availability of water from surface water and groundwater. And while there are opportunities to increase supplies to meet demands, we will see in the next few chapters that solutions to water scarcity will be driven by demand innovation and to a lesser degree increasing supplies.

Let's first examine the various sources of water.

Water sources

When I (Sarni) began my career as a hydrogeologist the need for water was always viewed as a supply issue with one answer: just find more water. As a result, I spent the early days of my career exploring for water for private and public sector clients and installing water supply wells.

For municipalities in the northeastern US, it was a straightforward solution: map the aquifers and install more wells. For multinational companies expanding into new areas, it was the same solution. There was never a question of managing demand.

Unfortunately, the supply solution focus still prevails. Typically during my presentations, when the projections of water demand versus water supplies are discussed inevitably the first couple of questions are: "Can't we find more freshwater? What about desalination?"

When we talk about water scarcity we are really referring to increased competition of water. For example, water demand in the US has tripled in the past 30 years while the population has only increased by 50 percent. Water consumption roughly doubles every 20 years globally, which is more than two times the rate of population growth.[1]

So let's break this down a bit. Projections of the gap between supply and demand do factor in some increase in water supply (Figure P1.2). Even with an increase in supply of about 20 percent (to 2030) there will be a shortfall of available water. Therefore, increasing supply is a *very* small part of the solution.

What about desalination as the answer to meet increasing demands? As we will see a bit later desalination is energy intensive and has geographic limitations (although brackish aquifers are being eyed for desalination). So, again this is only part of the solution.

What are the basics of water?

Water is in constant motion, and this cycle is the "hydrologic cycle," as illustrated in Figure 4.1.

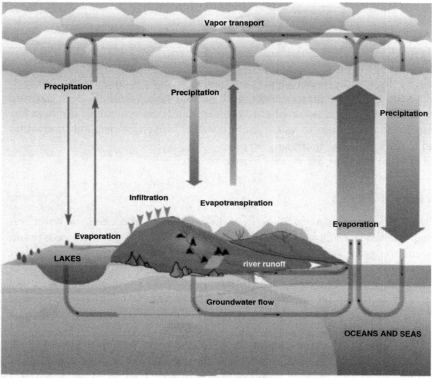

Note: The width of the arrows is approximately proportional to the volumes of transport water.

Figure 4.1 The general hydrologic cycle[2]

Freshwater evaporates from the oceans to form clouds, which float across the sky to deposit rain or snow on the land. Much of this freshwater eventually flows back to the ocean via streams, rivers, or storm drains, while a smaller percentage gets tied up in vegetation, buried in a glacier, or trapped in an underground aquifer until it is released back into the moving cycle via transpiration, snow melt, or the running of your garden hose, manufacturing processes, and so on.

This water cycle has not only been functioning effectively for eons (except the garden hose part), but has no identifiable beginning or end. Simply stated, water molecules don't ever disappear. They may change state, move from place to place, or get incorporated into more complex molecules, but the water balance on this planet has stayed relatively constant for millions of years. In fact, almost to the drop the exact same amount of water has existed on Earth since the time of dinosaurs.

When looking at readily available freshwater sources, there are three main sources: rainwater, surface water and groundwater (excluding glaciers). Each of these has its own unique procurement and distribution challenges. Recognizing that water flows through and around the planet in a circular fashion – the global water cycle discussed above – and that industry tends to move in a linear fashion, we will consider water supply the "front" of the water flow as it meanders, or is pushed, from source to use to process end.

Of the water available on earth the actual percentage of freshwater that is accessible is less than 1 percent (Figure 4.2). Although technologies are available to change this balance (desalination of salt and brackish water to freshwater), the demand for freshwater is creating scarcity.

In the US the sources of water have remained relatively the same over the past several decades with about 65 percent from surface water and 20 percent from groundwater with the remaining water from saline sources (Figure 4.3).

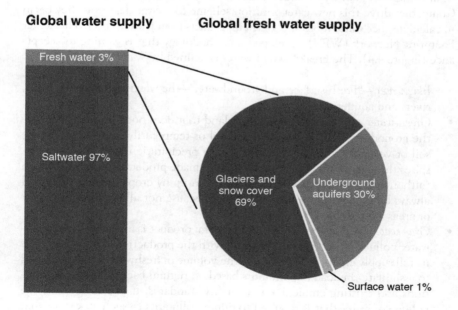

Global water supply   **Global fresh water supply**

Fresh water 3%

Saltwater 97%

Glaciers and snow cover 69%

Underground aquifers 30%

Surface water 1%

Figure 4.2 Global sources of water supply[3]

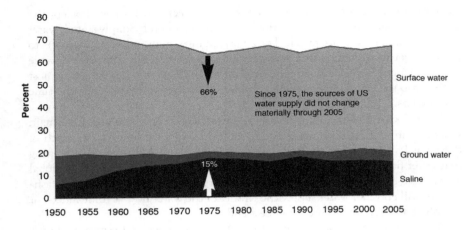

Figure 4.3 Change in sources of US water supply from 1950 to 2005[4]

While we will discuss water supplies in terms of rainwater, surface water and groundwater, it is worth noting that water is also being characterized in a somewhat different manner – broken down into blue, green and gray water. Companies drive this new categorization scheme to a large degree as they begin to calculate the "water footprint" of their products from jeans to beer. The Water Footprint Network (WFN) developed a methodology that is gaining in acceptance (Figure 4.4). The breakdown of water is defined by WFN as:

- *Blue water* – "Fresh surface and groundwater – the water in freshwater lakes, rivers and aquifers."
- *Green water* – "The precipitation on land that does not run off or recharge the groundwater but is stored in the soil or temporarily stays on top of the soil or vegetation. Eventually, this part of precipitation evaporates or transpires through plants. Green water can be made productive for crop growth (although not all green water can be taken up by crops because there will always be evaporation from the soil and because not all periods of the year or areas are suitable for crop growth)."
- *Gray water* – "The gray water footprint of a product is an indicator of freshwater pollution that can be associated with the production of a product over its full supply chain. It is defined as the volume of freshwater that is required to assimilate the load of pollutants based on natural background concentrations and existing ambient water quality standards. It is calculated as the volume of water that is required to dilute pollutants to such an extent that the quality of the water remains above agreed water quality standards."[5]

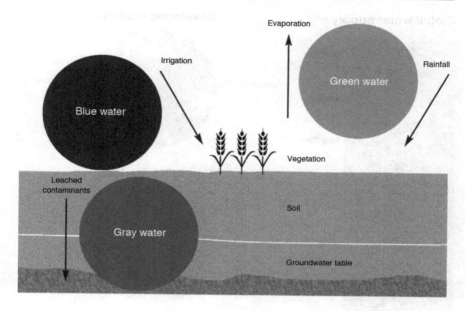

Figure 4.4 Types of water according to the Water Footprint Network[6]

It is now common for the public sector and private sector to refer to their blue, green and gray water footprint in addition to distinguishing between rainwater, surface water and groundwater.

How water is used is broken into three major categories: agricultural, industrial and residential/domestic (Figure 4.5). The relative water use in each category varies depending upon whether we are in a developed market/economy or emerging, as illustrated below. This is important as it is part of the puzzle in understanding technology opportunities and markets.

Before we go too far into this discussion, it is also worth a brief mention of water rights as it provides insight into one of the challenges (and perhaps opportunities) in driving water innovation on the supply side of the equation. Basically, who owns water and under what type of ownership law can be a challenge in promoting water innovation (rainwater capture in the western US states, for example).

Water in the US is divided into appropriative and riparian water rights (Figure 4.6). These two systems are fundamentally different: riparian water rights (based on land ownership) were derived from English common law and are common in the eastern states; prior appropriation rights, also known as the "Colorado Doctrine" of water law, derived from the maxim "first in time, first in right" and are common in the western US (Box 4.1). Each state has its own variations on these basic principles as informed by custom, culture, geography, legislation and

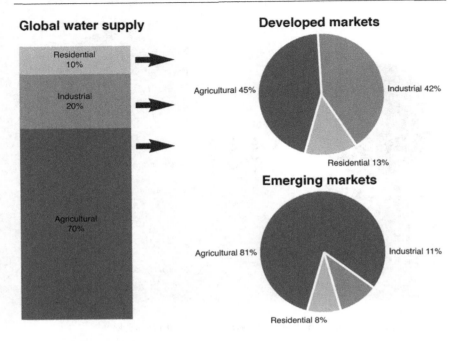

Figure 4.5 Global profiles of water uses[7]

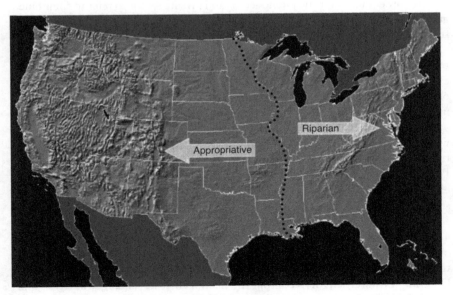

Figure 4.6 Water ownership profile of the United States – riparian versus appropriative rights[8]

case law.[9] The differences between state law makes the establishment of a national water policy challenging at best.

While the legal frameworks for water ownership can be a challenge in adopting new water technologies they may also represent an opportunity for market-based solutions to meet increasing water needs. A few thoughts on opportunities: there are increasing initiatives to establish agreements in managing water (in the western US); also, increasing investments in water rights beyond traditional ownership by farmers and ranchers (in the western US); and water rights as an investment (investment funds have about $300 million in water rights and it is projected to grow to $1.5 billion by 2013.[10]

Box 4.1 Prior appropriation doctrine and riparian doctrine

Water rights in the western United States operate under a prior appropriation basis, otherwise known as "first come, first own." This system means if a person first lays claim to a track of land, that person can use as much water from the wells, rivers, streams as he or she likes – in perpetuity. Along comes the next person in line, and they can use as much as they want of what's left, and so forth.

In the eastern part of the US, however, riparian doctrine holds. This system, based on English common law, states that an allotment of sorts should be used to make water available, as well as fair and accessible, to people whose property is adjacent to the source; the rights of one homeowner or landowner is weighed against the rights of all.

So back to rainwater, surface water, and groundwater to frame what we have in the way of supplies.

Rainwater and atmospheric water

It is estimated that the atmosphere has about 10.5 billion-acre feet of water or roughly about six times the amount of water in the world's rivers.[11]

While typically we do not think of atmospheric water as a supply side opportunity, increasingly technologies are being developed to extract moisture form the air for disaster response and military applications. Water moisture is essentially concentrated and used for drinking water supply (more on this later).

More obvious is the capture of rainwater (rainwater harvesting), which can be a clean, easily accessible drinking source. According to the NGO WaterAid, the term "rainwater harvesting" means "the immediate collection of rainwater running off surfaces upon which it has fallen directly." Rainwater harvesting is one of the oldest and most common forms of freshwater procurement used around the world. The capturing of rainwater for agricultural use and drinking in hand-dug rock catchment systems has been around for centuries.[12]

Today there is a movement to modernize the catchment and storage systems in places like Brazil and to introduce rain catchment where it has not been previously used. Most of these systems are built upon the basic concepts of capture, diversion and storage.[13] As discussed in an article by Richard Lucera,[14] not all areas are equal when it comes to the amount or continuity of rainfall, and long-term environmental impacts to areas where the captured rainwater would have traveled are relatively unknown. These are perhaps the biggest challenges facing effective rainwater harvesting, alongside ensuring hygienic storage for water designated for potable uses. In areas where other freshwater sources are scarce or depleted, however, rethinking rainwater capture is, and will continue to be, essential.

Surface water

Surface water is water accessed from rivers, lakes, streams, and runoff from rain. In many parts of the world, surface water is the main supply of freshwater. According to the United States Geological Survey, in 2005 about 80 percent of the water used nationwide originated from surface water sources.[15]

One of the challenges of surface water (and groundwater) is that it can be easily contaminated resulting in a further reduction of available freshwater for use. Once rainwater hits the ground, the likelihood of contamination from both point source (contaminants that enter the water way through a specific, obvious portal such as a pipe) and non-point source (diffuse contamination from various non-distinct sources, such as oil and fertilizer runoff from roads and lawns) increases. Though regulation of point-source pollution from large sources has been substantially mitigated, small-scale sources are more difficult to regulate.

Non-point source contamination is not easily regulated, as it is often the cumulative effect from multiple sources of pollution, usually scattered across a large area (Box 4.2). To tackle non-point source pollution is to tackle individual behavioral change, as well as large scale practices such as farming. Both continue to be very real impacts to the quality of surface water sources around the world. Here's why: people pollute water in myriad ways (e.g. sewage), and farms spill fertilizers and pesticides that contaminate water supplies.

According to the World Resources Institute, agriculture is the biggest source of nutrient and pesticide runoff into lakes and rivers in the US. Coastal areas fare no better. The journal BioScience reports that about 74 percent of the nitrogen discharged into the Gulf of Mexico comes from agricultural fertilizers.[16] These spills and runoffs create "dead zones," depleting organic life – lots of dead fish.

Stormwater runoff is another major surface water issue, especially where porous land has been covered with non-porous surfaces such as houses, pavement, and blacktop. By covering otherwise permeable surfaces that may have absorbed rainfall in reasonable time, more water is pushed at a faster rate into stormwater sewer systems, rivers, lakes and other bodies of water. This can result in greater contamination of waterways, including exposed drinking water reservoirs, as stormwater sewers are generally pipes leading directly to bodies of water

Box 4.2   Non-point pollution

On a small island village in Alaska near the Arctic Circle, Inuit people were becoming sick. The outbreak, it was discovered, was due to "splashing."

As an island village, Shishmaref was lacking in freshwater resources; a typical island fate. Therefore people were reliant on rainwater capture. The simple system many homes employed was to direct gutters to barrels, storing the freshwater or melted snow this way.

However, as the culture embraced some of the trappings of modernity, problems became apparent. All-terrain vehicles (ATVs) were used to transport sewage to the far end of the island. As the dirt roads were bumpy, sewage would spill, ending up in puddles. Splashes from the puddles would reach open gutters and infect the captured rainwater that was used for drinking and washing – with no further treatment.

Variances of these types of infective processes exist all over the developing world as people struggle to cope with increasingly concentrated populations and the need to develop water and sewage systems.

New water technologies can help prevent disease in many ways, from safe rainwater capture systems to wastewater management and treatment. Cost, of course, is a factor. But with populations concentrating in smaller areas and in bigger numbers than at any other time in history, the opportunity for proper water management systems is growing.

with no treatment or filtration.[17] Moreover, stormwater creates a greater risk of urban flash floods, and the salinization of freshwater sources from non-point source pollution[18] and from the increased flow of freshwater to saltwater.[19] These further contaminate our supply of freshwater.

Groundwater

As the name suggests, groundwater is water beneath the land surface. Water below land surface is divided into the unsaturated zone and saturated zones: the unsaturated zone has water in the pore spaces of the soil and rocks but does not completely fill the open pore spaces, and the saturated zone contains water in 100 percent of the pore spaces. It is in the saturated zones (aquifers) where we extract water for agricultural, commercial and domestic uses. In general, aquifers are categorized as unconfined and confined aquifers. Unconfined aquifers are also referred to as water table aquifers – one can drill a well and tap into water immediately below the land surface. Confined aquifers are water at depth and under pressure – when tapped, water will rise above the top of the aquifer and in some cases above land surface.

In general, groundwater supplies are dwindling. Representing roughly 30 percent of the global freshwater supply on the planet, these water sources often took millennia to fill through natural filtering and seepage from rainwater and surface water sources, and are now being drained at a rapid pace, mainly for agriculture and drinking water.[20] We are essentially "mining" groundwater – extraction far exceeds recharge of water into groundwater aquifers (Figure 4.7). Several areas around the world rely heavily on groundwater, such as the Eyre Peninsula of Australia, which gets approximately 85 percent of its water from underground,[21] and the US Midwestern Ogallala Aquifer.[22]

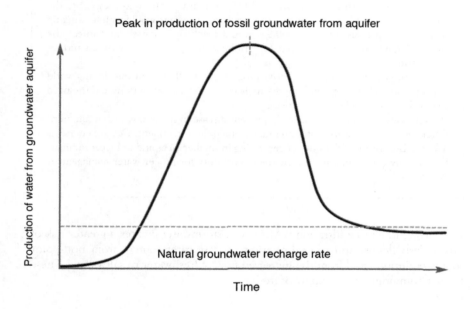

Figure 4.7 Potential peak water curve for fossil groundwater production[23]

The "mining" of groundwater impacts the supply side of the equation through reduced availability of freshwater and increased contamination.

The agricultural use of groundwater is a major contributor to the salinization of freshwater supplies. The lowering of groundwater levels can result in an increase in metals often caused by excess salt generated from agricultural practices and/or lead to increases in naturally occurring metals and other potential contaminates from surrounding soils.[24]

In addition to the reduced quantities of groundwater available for use, groundwater pumpage is resulting in subsidence and impacting cities such as Tucson (Arizona) and Mexico City.

While increasing supplies of water will not meet the projected needs of an increasing global population, there are technologies that are being developed to address at least part of our needs.

Technologies

There are really just a few options in increasing the supply of water to meet increasing demands; find more water through improved exploration technologies, improve the capture of water (rainwater and air moisture), and improve the management of surface water and groundwater. Some of these are focused on the application of fairly simple technologies (rainwater capture) or more sophisticated technologies such as remote sensing to identify previously undiscovered aquifers.

Rainwater and atmospheric water

Existing rainwater harvesting technology is focused mainly on the use of appropriate (non-toxic and easy to clean) materials for rooftops and other surfaces and storage, the latter being the more costly of the two. In Singapore harvesting of rain is currently being utilized on both high-rise residences and in a larger scale system at the Changi Airport, which includes runoff from runways as well as buildings. In Germany a few large-scale harvesting systems have been designed into urban renewal projects, allowing for water conservation as well as a reduction in potential pollution from sewage overflow due to storms. Both of these rainwater capture examples focus on non-potable use for the captured water, such as toilet flushing, air conditioning, and in some cases, garden water. In other parts of the world, such as the Gansu Province in China, Bermuda and Hawaii, the rainwater captured in systems both large and small in size is used directly for drinking water.[25]

The types of storage vessels for captured rainwater vary greatly around the world. In the United States, polyethylene tanks are the most common used. In developing countries, a material called "ferrocement," which is made of a steel and mortar composite, is common because of its low cost and ease of use. Most rainwater storage tanks tend to be constructed of locally available materials and will have to take into account many local factors, such as daily temperatures, consistent rainwater supply, demand, and aesthetics.[26]

Deciding whether the end use is for potable or non-potable use is also an important factor in the choice of construction materials. However, there are several places around the world where open pits and even swimming pools are used as catchment storage regardless of end use. Clearly, there is room for innovation with storage, but rainwater capture technologies are increasing.

There are some interesting new takes on the age-old practice of rainwater harvesting. In Prescott, Arizona there is a discussion about the feasibility of large-scale rainwater harvesting (LSRH). In a semi-arid climate where there is heavy use (and overdraft) of groundwater, implementing LSRH could reduce overdraft of groundwater by an estimated 11,000 acre-feet annually, which is about the same amount a town the size of Prescott uses in one year. The LSRH being examined in Prescott involves three different methods: the use of existing steep mountain grades to direct and capture water; the removal of vegetation and physically altering flatter grounds; and collection from urban surfaces such as rooftops and streets. While the main focus is reducing the need to draw from the aquifer, there is also discussion of using the rainwater to recharge the aquifer.[27]

Rainwater is generally seen as clean and a good source for drinking water; however, there are emerging conversations around pushing the limits further into "water neutral" buildings. These new buildings can utilize rainwater for non-potable means, such as toilet flushing and landscape water, as well as potable means (drinking).

Kroon Hall at Yale University (New Haven, Connecticut, US) took this approach in 2007 while going for LEED Platinum status within the US Green Building Council LEED certification. Because treated potable water is used for the activities mentioned above, the use of rain catchment from the rooftop and grounds meant a savings of half a million gallons of potable water annually with a 10-year payback period.[28] The green building councils of both Australia and the US are promoting water neutral buildings, such as the Pixel building in Melbourne, Australia, where rainwater harvesting is used in several ways, including watering the vegetated roof.

An example of innovation in rainwater harvesting is the "Rainwater HOG" system.[29] Rainwater HOG was the runner up in the 2010 Imagine H2O annual water technology innovation competition.[30] The Rainwater HOG system uses modular tanks to catch and store rainwater for reuse within residential applications such as irrigation or showering. The rainwater HOG systems are manufactured in the US and Australia and have gained increasing acceptance in the building sector.

As previously discussed, atmospheric water can also be harvested for use. A winner of the 2011 James Dyson Award was a technology that mimics the desert Namib beetle. Edward Linacre of Swinburne University of Technology in Melbourne, Australia created an irrigation system that can pull liquid moisture out of dry desert air. "Airdrop", as the system is known, borrows from the Namib beetle, which can live in areas that receive just half an inch of rain per year by harvesting the moisture from the air that condenses on its back during the early morning hours. A hydrophilic skin helps to capture water molecules from the dry air and then accumulates them into droplets of consumable liquid water.

The self-powering device pumps water into a network of underground pipes, where it cools enough for water to condensate. From there the moisture is delivered to the roots of nearby plants. Linacre's math shows that about 11.5 milliliters can be harvested from every cubic meter of air, and further development could raise that number even higher.

Such a system could provide regular moisture to plants being grown in the world's driest regions. And because it is low cost and self-powered, there's not a lot of investment or maintenance involved in deploying Airdrop.[31]

Surface water

The most common technologies surrounding surface water use have to do with diversion and retention. Perhaps the most known combination of technologies are dams, ditches and reservoirs, whereby surface water is backed up at its site of origin to increase volume, diverted from its original location, and then stored in an open, fabricated lake for future use. Several controversial dam and storage systems have been built, some for hydro-electric power such as the Three Gorges Dam in China and Hoover Dam in Nevada, and some for water diversion such as the Central Arizona Project, which brings roughly 1.5 million acre-feet of water through a series of tunnels, pipes, pumping pants and aqueducts from the Colorado River to various Arizona counties where water demand is higher than local resource availability allows.[32] Regardless of designated use of the dam and diversion system, there can be major environmental effects, most notably, in-stream flow, nutrient distribution, fish and aquatic species movement, and the physical stream degradation.[33]

Rapid evaporation is also a concern surrounding open water storage, especially considering the increased temperatures the world is experiencing due to climate change.[34] This is a larger concern in arid and semi-arid climates, which makes up a not-insignificant 50 percent of the earth's surface.[35]

Even technologies to limit evaporation are being developed. One example of technology innovation to address this issue is More Aqua Inc. (MAI), an MIT-based start-up addressing the need to suppress the evaporation of freshwater from constructed reservoirs.[36]

According to MAI:

> in a typical dry-climate reservoir such as in Texas or Australia, the evaporation losses exceed the usage rates for domestic and industrial uses (excluding irrigation). For example, 42 percent of the surface water supply of Texas is lost to evaporation. These enormous losses are being offset by massive reservoir construction projects, desalination plants, and ever longer pipelines, all at enormous economic and environmental cost.[37]

MAI's technology consists of applying a monolayer film (an environmentally-benign material extracted from vegetable oil to the water surface), which can reduce evaporation up to 80 percent according to MAI. The film acts as a diffusion barrier for water vapor but not for oxygen and carbon dioxide.

MAI's technology controls the film in conditions of high wind, turbulence and other effects, which have precluded success with this technology in the past. As of the date of this book, the technology was proven in the laboratory and ready for pilot testing according to MAI.

While common, open reservoirs do not represent leading-edge innovation opportunities in the water sector, but they do meet large-scale demand needs reliably.

A common use of water, landscape water management, is seeing the application of innovative technologies. By employing the idea of landscape design features such as bioswales and rain gardens, storm water runoff can be held in areas where the plants and soil are able to utilize the water until it evaporates and percolates into the groundwater. By utilizing indigenous plants that live on the edge of wetlands and other areas of naturally intermittent water flow, these areas also increase habitat for local fauna. In heavily populated areas these design features can be scattered about without altering existing infrastructure, while also adding much needed pockets of greenery. Features like these can also be incorporated in greater number in areas where there is already a fair amount of green.

In Queens (New York), the Queens Botanical Garden created various bioswales throughout their grounds in 2007. The bioswales not only help ease storm water runoff issues, but have also become an educational tool for discussing water issues in the city.[38]

Groundwater

Groundwater is generally accessed through wells that are installed by drilling or boring technologies. There are a variety of drilling technologies used around the world. In the poorer regions of the world, UNICEF is working to promote manual well drilling as a more cost-effective and less damaging means of accessing groundwater. UNICEF has also created a series of Groundwater Programming Principles, along with a Code of Practice for Cost-Effective Boreholes, and is dedicated to raising awareness about the fragility of groundwater sources.[39] Commonly used technologies in other parts of the world are traditional auger drilling, sonic drilling, and direct push technology.[40] There are some innovations in well drilling technologies but not a focus area of water tech innovation.

Aquifer recharge is increasingly being used as a means to replenish groundwater resources. In arid areas where groundwater is extracted faster than natural recharge rates, aquifer recharge is a viable approach to a strategic water management plan. Two common ways of recharging groundwater sources are infiltration basins and injection wells. Infiltration basins are essentially small human-made ponds or streambeds that allow water to percolate through the soil and into the aquifer. The natural percolation process has the added benefit of removing many contaminates from the surface and/or rainwater. This technology is used globally, from California in the US to Zimbabwe, and is often used in conjunction with storm water management plans as a means to capture runoff.

Injection wells are deeper wells and are often created in conjunction with a surface water storage system, and can even be constructed from existing wells that have gone dry.[41]

Remote sensing technologies can also meet an important need in increasing access to groundwater resources in remote areas of the world. It is difficult to explore for groundwater in large unpopulated areas. As a result, aerial and increasingly satellite imagery are being used to identify potential areas for groundwater exploration.

Satellite remote sensing provides a valuable global overview that can monitor changes in rainfall, extent of water bodies, vegetation and – at a more local level – help to identify zones with groundwater potential. As previously discussed, groundwater plays an important role; it feeds springs and streams, supports wetlands, and maintains land surface stability. Groundwater is a significant global resource, comprising 96 percent of the Earth's unfrozen freshwater, as well as being the main water source in many water-scarce areas.[42]

As a result, organizations such as UNESCO under the International Hydrological Programme (IHP) and the European Space Agency have cooperated with other space agencies, UN organizations and African partner organizations, to use satellite data within the framework of the TIGER project, "to develop a network of experts to strengthen the scientific base and work towards developing sustainable satellite-based information services to support water resources management." UNESCO's IHP provides methodological and technical advice for the better management of these groundwater resources through a series of projects.[43]

The Desert Research Institute (DRI), located in Las Vegas (Nevada), is developing data interpretation strategies that integrate various data types to characterize groundwater resources for the identification of exploratory well locations. In many developing countries, readily available remote sensing data may comprise a majority of the existing information over local and regional areas and these data can be used in maximizing water development efficiencies. These strategies are being developed using a Geographic Information System (GIS) as the unifying element for all collected data and will be tested in conjunction with an ongoing World Vision International (WVI) drilling project in Ghana, West Africa. The steps undertaken during the collection and integration of data from literature, maps, borehole records, Global Positioning System (GPS) receivers, field observations, and several remote sensing platforms are essential in creating a GIS-based hydrogeologic model of the study area adequate to support strategy development and testing. For example, these techniques are being used in Ghana to select locations for exploratory groundwater wells with the highest probability of success.[44]

NASA's twin satellite program, the Gravity Recovery and Climate Experiment, or GRACE, has also become a valuable asset in understanding groundwater extraction.[45] According to NASA, GRACE "provides user-friendly data grids, with most corrections applied, to analyze changes in the mass of the Earth's hydrologic components."[46]

The search for additional groundwater resources continues.

In a recent article in *Environmental Research Letters*, MacDonald and Taylor noted that there is little quantitative information on groundwater resources in

Africa.[47] Through their research, they now estimate that there is about 0.66 million km³ (0.36 to 1.75 million km³) in groundwater storage of which not all is available. This estimate is more than 100 times estimates of annual renewable freshwater resources in Africa. The largest volumes are in the North African countries of Libya, Algeria, Egypt, and Sudan.

While the new estimates of groundwater resources are encouraging the authors point out that small hand pump wells for abstraction contain adequate storage for variations in recharge. In contrast, higher yield wells are more limited. As a result strategies for increasing irrigation or supplying water to rapidly urbanizing cities predicated on the widespread drilling of high yielding wells are likely to be unsuccessful.

This recent mapping is a starting point. According to MacDonald and Taylor:

> As groundwater is the largest and most widely distributed store of freshwater in Africa, the quantitative maps are intended to lead to more realistic assessments of water security and water stress, and to promote a more quantitative approach to mapping of groundwater resources at national and regional level.[48]

Before we move on to the demand side of the equation, it is important to recognize that there are opportunities to jointly develop surface water and groundwater resources together. The extraction of surface water and groundwater strategically is referred to as "conjunctive use" and innovation opportunities exist in sustainable water management. While not technology-focused, conjunctive use can be part of a water strategy, which includes technology, and water governance approaches.

The goal of conjunctive use includes extraction from groundwater aquifers to "buffer" water supply availability against the high flow variability and drought propensity of many surface watercourses. This approach is important for the mitigation of climate change impacts, manifested as persistent droughts, and can be one of the best ways to confront some of the serious problems of groundwater salinization and soil water logging on alluvial plains.[49]

Now on to a discussion of the demand side of the equation – efficiency, recycling, and reuse – where the real opportunities to address sustainable water management reside.

Notes

1   CitiGroup (2011) *Citi Water Sector Handbook*, May 24.
2   Adapted from UNEP and from William Sarni (2011) *Corporate Water Strategies*, London: Earthscan.
3   Adapted from CitiGroup (2011), *op. cit.* note 1.
4   Ibid.
5   www.waterfootprint.org.
6   Sarni (2011), *op. cit.* note 2.

7   Adapted from CitiGroup (2011), *op. cit.* note 1.
8   Ibid.
9   http://en.wikipedia.org/wiki/Water_right.
10   CitiGroup (2011), *op. cit.* note 1.
11   Ibid.
12   www.wateraid.org/international/what_we_do/sustainable_technologies/technology_notes/2055.asp.
13   www.unep.or.jp/ietc/publications/urban/urbanenv-2/9.asp.
14   Richard Lucera (2011) "Exploring the Feasibility of Rainwater Harvesting in Southern California," *Stormwater: The Journal for Surface Water Quality Professionals*, 30 April, www.stormh2o.com/SW/Articles/14217.aspx.
15   http://ga.water.usgs.gov/edu/wusw.html.
16   www.ewg.org/agmag/farm-subsidies-lead-to-ocean-pollution-researchers-say.
17   http://water.epa.gov/action/weatherchannel/stormwater.cfm.
18   www.learner.org/courses/envsci/unit/text.php?unit=8&secNum=7.
19   www.climate.org/topics/water.html.
20   www.sciencenews.org/view/generic/id/337097/title/Groundwater_dropping_globally.
21   www.epnrm.sa.gov.au/Water/Groundwater/Introduction.aspx.
22   http://environment.nationalgeographic.com/environment/freshwater/groundwater/#.
23   Sarni (2011), *op. cit.* note 2.
24   www.unep.org/dewa/Portals/67/pdf/Groundwater_INC_cover.pdf.
25   www.unep.or.jp/ietc/publications/urban/urbanenv-2/9.asp.
26   www.harvesth2o.com/rainwaterstorage.shtml.
27   www.dcourier.com/main.asp?SectionID=36&SubsectionID=73&ArticleID=93031.
28   http://news.yale.edu/2010/02/01/kroon-hall-achieves-leed-platinum-certification.
29   http://rainwaterhog.com.
30   www.imagineh2o.org.
31   www.jamesdysonaward.org/Projects/Project.aspx?ID=1722.
32   www.cap-az.com.
33   www.internationalrivers.org/node/1545.
34   www.fao.org/docrep/V5400E/v5400e0c.htm.
35   www.jstor.org/pss/4312240.
36   www.moreaqua.com.
37   Ibid.
38   www.queensbotanical.org/103498/sustainable/bioswales   and www.sustainablecitynetwork.com/topic_channels/environmental/article_19b62066-8f24-11e0-8511-0019bb30f31a.html.
39   www.unicef.org/wash/index_49090.html.
40   www.groundwaterprotection.com/services.html.
41   http://water.epa.gov/infrastructure/drinkingwater/sourcewater/protection/index.cfm.
42   www.globalwaterforum.org/2012/02/13/non-sustainable-groundwater-sustaining-irrigation.
43   http://portal.unesco.org/science/en/ev.php-URL_ID=6453&URL_DO=DO_TOPIC&URL_SECTION=201.html.
44   Timothy B. Minor, Jerome A. Carter, Matthew M. Chesley and Robert B. Knowles (1994) *An Integrated Approach to Groundwater Exploration in Developing Countries Using GIS and Remote Sensing*, Alexandria, VA: Topographic Engineering Center.
45   www.globalwaterforum.org/2012/02/13/non-sustainable-groundwater-sustaining-irrigation.
46   http://grace.jpl.nasa.gov/information.
47   April A. M. MacDonald and R. G. Taylor (2012) "Quantitative Maps of Groundwater Resources in Africa," *Environmental Research Letters*, 19 April.
48   Ibid.

49 Stephen Foster, Frank van Steenbergen, Javier Zuleta and Héctor Garduño (2010) *Sustainable Groundwater Management: Conjunctive Use of Groundwater and Surface Water from Spontaneous Coping Strategy to Adaptive Resource Management*, Contributions to Policy Promotion Strategic Overview Series, no. 2, World Bank, South Asia Region.

Chapter 5

Water demand

Who needs water? The agricultural sector? The commercial sector? The retail sector? What about the ecological systems upon which we depend? And how do we develop an integrated approach to water management and demand?

At the forefront of developing integrated water management strategies is the World Wide Fund for Nature (WWF, along with other NGOs, academic institutions, the public sector and companies). Stuart Orr, head of water stewardship for WWF International in Gland, Switzerland, offers some insight into how water scarcity can be understood through risk:

> Water risks affect people in different ways and are highly specific, involving the loss of certain fine-grained and often crucial information about the risk to specific stakeholders. Highlighting and explaining how securing adequate water supplies leads to major risk reduction, means engaging the diverse set of stakeholders who benefit from water use, and are often, unwittingly vulnerable, to its scarcity.
>
> Aggregated water scarcity maps are useful for raising awareness, but reveal little about the implications of water scarcity for people and ecosystems. How does one go about identifying the regions and stakeholders that are most likely to be affected by shortages? How do we begin to understand the problems that water scarce regions and their people are likely to confront? The problem can only be understood, and action can only be effectively tailored, by focusing at the local level. Global and national scale macro assessments do not reveal the dynamics, components and the biases within the water scarcity phenomenon.
>
> While almost everyone recognizes water scarcity to be a "public bad", the manner in which water scarcity impacts government and business through complex social and ecological systems is less well understood. What is more certain are that threats to people and all they value, necessitates a determined approach to exploring risks associated with water management failures. Water scarcity is caused more by the nature of demand and the allocation of water than the general availability of water – an issue best addressed through better water management and governance. As such, this

is less a resource crisis and more often an issue caused by policy and institutional failure rather than by technical failure.

Identifying the nature and location of future water scarcity risks requires imagination as well as knowledge. Risks associated with water scarcity can be classified as follows:

(a) Risk from insufficient water resources to meet the basic needs of people, the environment and business, which in turn leads to...

(b) Risk from the consequences of insufficient water resources, such as higher energy prices, loss of competitive advantage, political and economic instability, biodiversity loss and ecosystem breakdown, population migration, or lost economic opportunities to name a few; and as a result...

(c) Risk from poor water management decisions taken in reaction to water scarcity, with negative consequences for some or all users. Such decisions may be a result of political or economic expediency, short-term thinking, lack of knowledge or capacity or simply desperation and lack of choice.[1]

We know that agriculture currently uses on average 70 percent of global freshwater withdrawals, indicating that the world's demand for freshwater water is determined predominantly by its demand for food. Population growth, increasing affluence of emerging economies and over-consumption in developed economies will likely lead to a doubling of the demand for agricultural production by 2050. The food we eat is becoming more water-intensive, and different types of agriculture use different amounts of water. It takes 1,000 liters of water to create a kilogram of wheat, but it takes 7 to 15 times as much to create a kilogram of beef because of the feed-grains involved. As the world becomes wealthier and people can afford to introduce more meat into their diets, the water demands of agriculture and particularly for feed stock will become greater.

In the past 50 years, water use trebled as the world's population increased by 3.5 billion people. The challenge will be supporting 3 billion more, all of whom will need to be fed and watered. In the short term, over-abstraction of water is a widespread water resources problem, but in the longer term anthropogenic climate change and associated aridity will exacerbate problems caused by inadequate water mismanagement. It is this mismanagement which will undermine crucial environmental capacity and resilience that would otherwise have aided adaptation to climate change. Reduced water runoff and aquifer recharge is expected to lessen total available water in parts of the Mediterranean basin, southern Africa, Australia, south and southeast Asia, and the Americas.

Currently, rain-fed agriculture accounts for about 60 percent of crop production, but because it relies on direct rainfall it rarely takes water away from other competing interests. As lands receive less direct rainfall due to climate changes, farmers are confronted by choices which might include abandoning certain crop production, a shift of crop type or supplementation of water through irrigation.

Demographic shifts and simultaneous economic growth require increased amounts of water for urban households, energy and industry. This sets up a direct competition for water between rural agriculture and urban growth, which the latter will generally win. However, the real losers in this type of competition are the social and environmental interests with less political and economic agency that depend on functioning river ecosystems.

The increased strain on the environment and higher demand for food and the growth in bio-fuels has already been reflected in higher soft commodity prices over the first decade of the twenty-first century. This has attracted hedge funds and speculators into the agricultural sector. Where land is unproductive, this external investment can be of benefit to rural economies by provision of capital, machinery and expertise to develop agricultural land that would otherwise be left unproductive. However, these investments can also supersede existing rights and laws for relatively voiceless farmers and communities. In addition, rapidly increasing commodity prices have a direct knock-on effect on food security (particularly for the poorest) and the supply chains and profitability of local producers.

When translated into domestic policy decisions these economic pressures may contribute to unforeseen political risks. Protectionist policies may be popular at a national level, but lead to unstable prices and inefficient allocation of resources at a regional or global level.

When faced with more unstable policy decision, producers dependent on commodities for their supply chains may experience conflicting needs, a stable domestic economic and policy environment in order to invest in their local productive capacity, and access to international markets to sell their goods. Clearly, although agriculture's short-term water needs could be partly satisfied at the cost of environmental and other interests, it is likely that there will be substantially less water available globally for agricultural production in the long term. While some of this may be counteracted by improving agricultural technology and water use efficiency, politically relevant and economically viable ways of allocating water to and within agriculture will have to be adopted to ensure equitable, sustainable and efficient management of water resources at a basin scale.

It is also critical to recognize the central nature of water as a catalyst or constraint on development and that the water discourse needs to be deeply embedded in national and international dialogues around agriculture, energy, trade and infrastructure finance. Inadequate attention to these linkages may result in perverse outcomes in the relationship between agricultural commodities, land and water, with the associated risks for food security and supply chains. As an example, one concern is that the markets and prices for commodities with embedded water are largely unregulated and this may become the most contested domain for water. As it becomes clearer that water markets will not be established to enable speculation and significant profit generation in bulk water, investment focus will shift to where water is manageable and profitable; trade in commodities with significant amounts of embedded water as well as the means of their production."

The challenge of reducing demand

Part of the solution to water scarcity is reducing demand for water through vastly increased efficiency and reduced losses during the transport of water (leakage). The challenge in increased efficiency and reducing leakage is tied to the current cost of water.

As it stands now, the price of water really doesn't promote efficiency and conservation.

As previously discussed, we need to distinguish between the cost of water, the price of water and the value of water. The current price of water doesn't fully capture the sunk costs for infrastructure and the delivery (the costs of energy to extract, move and treat water) of water. If the current price did capture all of these costs we would have the full cost of water. If we considered the real value of water in terms of business continuity, brand value, and so on, we could quantify the full value of water. In general, businesses are moving towards recognizing the full value of water.

It is worth digging into the issue of water pricing a bit more, as it is *the* barrier to the efficient use of water. Water can be had in many areas cheaply or essentially free of charge. Without financial consequence, the incentive for change becomes a purely moral enticement. This is why there is a strong and growing movement by people who believe water needs to be priced unilaterally. Once there is a price for water, the theory goes, the penalty for waste and inefficiency can be tallied.

According to the International Finance Corporation (IFC), one of the first things that should be done to alleviate the water crisis is to put a price on water.[2] The IFC believes this can be a tool for better water management.

The OECD promotes policies that improve the social and economic well-being of people around the world. One of the items identified as most pressing is water. The reason the OECD gives is increasing competition between water users to access the resource.[3] The OECD believes fair pricing will drive solutions to access to water for all users.

Several OECD studies propose that the "right price" for water will encourage people to waste less, pollute less, and invest more in water infrastructure.[4]

As noted, prices vary significantly: comparable volumes of water in Denmark and Scotland can cost 10 times more than in Mexico while Irish households pay no direct fees for water. Based upon the OECD studies, increases in water bills over the past decade were primarily driven by higher wastewater charges to cover the costs of investment in environmentally sound treatment and disposal. In many OECD countries it now costs more to get rid of wastewater than to bring in drinking water, the organization notes.[5]

The global perspective on water pricing is illustrated in Figure 5.1.

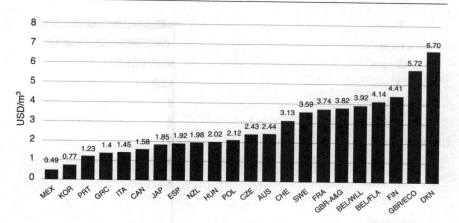

Figure 5.1 Unit price of water supply and sanitation services to households (includes taxes; US$/m³)[6]

These costs are all built into the price of water: extraction, transportation, treatment, management, and wastewater. But there is no single formula. For example, people living on low incomes in Hungary and Mexico sometimes pay over 4 percent of their disposable income on water and wastewater services (Figure 5.2). And that is just on the "people" side of things – the domestic portion of water use. When agriculture is woven into the equation, the supply and demand side changes further (Figure 5.3).

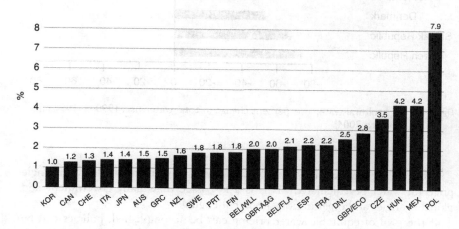

Figure 5.2 Water supply and sanitation bills as a share of disposable income: average income of the lowest decile of the population[7]

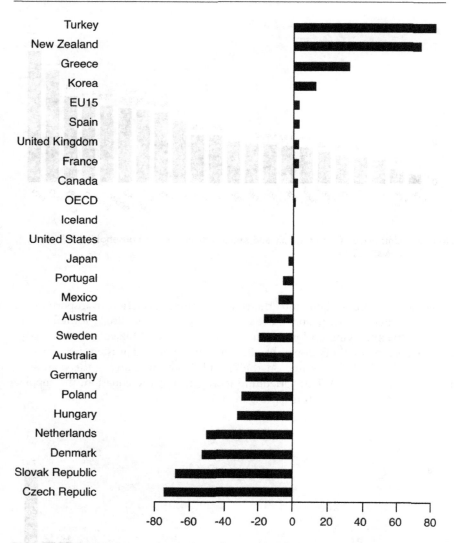

Figure 5.3 Agricultural water use: percent change in total water use, 1990 to 1992 and 2002 to 2004[8]

So, with varying demands, supplies, and uses, how in the world can a single price for water be calculated? This is a key issue that many government and global agencies and organizations find challenging.

If the goal of equitable water pricing can be accomplished, policies can be developed to allow for government pressure and mandates. As we have seen, artificially keeping water prices low has consequences.

The OECD also says government subsidies for agricultural production can often encourage wasteful water use and pollution. It says that in some countries lower agricultural subsidies – which would in effect raise prices, including for water and energy, are making farms cleaner and more efficient. In many developing countries increasing investment in water and sanitation infrastructure is a priority, albeit one that exists more on paper than in actionable policies or programs.

Moreover, the world's poorest people typically pay 5 to 10 times more per unit of water than do people with access to piped water.[9]

A new way of viewing how to strengthen investment in water and sanitation services is provided in the OECD study *Innovative Financing Mechanisms for the Water Sector*.[10] It provides examples of successful financial models such as the Indian state of Tamil Nadu where improved access to capital markets by small waste utilities by pooling water and sanitation projects were turned into investment packages and combined different sources of capital to fund the packages. And where and when financing for water becomes robust, so too do the business and technologies that may attract such capital.

With agricultural production projected to double by 2050 to feed the growing world population, farmers will need to improve water efficiency. *Water: the right price can encourage efficiency and investment*, a report from the OECD, suggests that farmers should pay not only the operation and maintenance costs for water but also their fair share of the capital costs of water infrastructure. "In areas where the price of agriculture water has increased, agricultural production has not fallen – Australia managed to cut irrigation water use by half without loss of output. According to the report, the uptake of existing water-saving irrigation techniques in China and India, both large agriculture water users, is expected to help check the global agricultural water use to 2050. Trends in water withdrawals in the agricultural sector are illustrated in Figure 5.4.

Clearly there are many benefits to pricing water and the OECD is advancing this position. So are many other groups and organizations.

However, the equitable pricing of water is challenging and raises several key questions. Who owns water; should it be made part of the "commons" and be made free for all? How will the poor pay for water?

In a 2009 *Huffington Post* editorial titled "Misconceptions about Water Pricing," water pricing advocate Robert Stavins (director of the Harvard University environmental economics program) says it's worth tackling water even with its challenges.[11] According to Stavins:

> One misconception is that "because water prices are low, price cannot be used to manage demand." This misconception that low prices somehow obviate the use of price as an incentive for water conservation may stem from economists' definition of a price response in the range observed for water demand as "inelastic." There is a critical distinction between the technical term "inelastic demand" and the phrase "unresponsive to price."

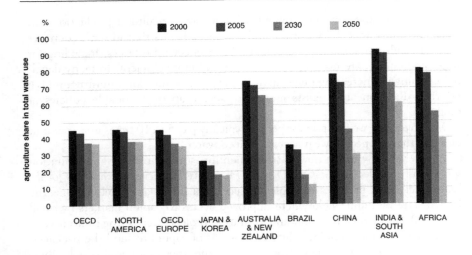

Figure 5.4 Water withdrawals from 2000 to 2050: share of agriculture in total water withdrawals[12]

Inelastic demand will decrease by less than one percent for every one percent increase in price. In contrast, if demand is truly unresponsive to price, the same quantity of water will be demanded at any price. This may be true in theory for a subsistence quantity of drinking water, but it has not been observed for water demand in general in 50 years of published empirical analysis.

A second misconception is that "water customers are unaware of prices, and therefore price cannot be used to manage demand." If this were true, the hundreds of statistical studies estimating the price elasticity of water demand would have found that effect to be zero. But this is not the case. Instead, consumers behave as if they are aware of water prices. The hundreds of studies reviewed by Stavins cover many decades of water demand research in cities that bill water customers monthly, every two months, quarterly, or annually; and in which bills provide everything from no information about prices, to very detailed information. His conclusion is that water suppliers need not change billing frequency or format to achieve water demand reductions from price increases, but providing more information may boost the impact of price changes.

A third misconception is that "increasing-block pricing provides an incentive for water conservation." Under increasing-block prices (IBPs), the price of a unit of water increases with the quantity consumed, based on a quantity threshold or set of thresholds. Many water utilities that have implemented IBPs consider them part of their approach to water conservation; and many state agencies and other entities recommend them as water conservation tools. But analysis indicates that increasing-block prices, per se, have no impact on the quantity of water demanded, controlling for price levels.

A fourth and final misconception is that "where water price increases are implemented, water demand will always fall." Price elasticity estimates measure the reduction in demand to be expected from a one percent increase in the marginal price of water, all else constant. Individual water utilities may increase prices and see demand rise subsequently due to population growth, changes in weather or climate, increases in average household income, or other factors. In these cases, a price increase can reduce the rate of growth in water demand to a level below what would have been observed if prices had remained constant.

Raising water prices (as with the elimination of any subsidy) can be politically difficult. This is probably one of the primary reasons why water demand management through non-price techniques is the overwhelmingly dominant approach in the United States. But the cost-effectiveness advantages of price-based approaches are clear, and there may be some political advantage to be gained by demonstrating these potential cost savings.[13]

Summing it all up, Angel Gurría, the OECD secretary-general, says, "Putting a price on water will make us aware of the scarcity and make us take better care of it."[14]

In view of the challenge of current water pricing and what is needed to change pricing there are emerging technology solutions to address the inefficient use of water and high rates of water leakage.

Let's explore the technology opportunities and what can be accomplished with increased efficiency and reduced leakage.

Efficiency and leakage

Increased water efficiency and reduced leakage will reduce stress on available freshwater supplies. This is no different than the benefits of energy efficiency and conservation on reducing the need for increases in power generation (with resultant benefits of reduced capital and operating cost requirements).

In 1989, Amory Lovins, chief scientist of the Rocky Mountain Institute,[15] coined the term "negawatt." Negawatt power is a theoretical unit of power representing an amount of energy (measured in watts) saved. The energy saved is a direct result of energy conservation or increased efficiency.[16]

The "negadrop" is a similar construct – water saved is less water required from freshwater sources such as surface water, groundwater and rainwater.

Water efficiency and conservation is taking hold – in particular in water "intensive" sectors. Actually, it is more like *water visible sectors* – where water is a significant and highly visible component of a product such as a beverage.

An example of this increasing focus on water efficiency is illustrated in the recent report by the Beverage Industry Environmental Roundtable (BIER).[17] The report, titled *Water Use Benchmarking in the Beverage Industry*, provides a summary of the annual water efficiency performance by BIER member companies.[18]

This fifth annual quantitative benchmark by BIER was performed to evaluate water use and efficiency performance across more than 1,600 beverage-manufacturing facilities. BIER undertakes this annual benchmarking study to more clearly understand water use drivers and impacts and to demonstrate improvement as an industry sector in water stewardship efforts. Members utilize the benchmark data to support their individual water conservation programs as needed:

> The study aggregates facility-specific data from each member company covering the past three years of operations and analyzes results by four distinct beverage production facilities: bottling, brewery, distillery and winery. Further, assessment of water risk and opportunities as well as some of the best practices employed to drive water use avoidance and efficiency are summarized.[19]

The results from the latest BIER study are summarized below.

- The industry aggregate water use ratio improved by 9 percent from 2008 to 2010.
- Approximately 69 percent of facilities improved their water use ratio from 2008 to 2010.
- Aggregate beverage production remained relatively stable, increasing 1 percent from 2008 to 2010 while industry aggregate water use decreased approximately 8 percent from 2008 to 2010.
- By improving water use efficiency, the industry avoided the use of approximately 39 billion liters of water in 2010 – enough water to supply the entire population of New York City for eight days.[20]

The beverage sector is not alone on focusing on water efficiency. Increasingly, companies are measuring their water footprint and looking at ways to improve water efficiency across their entire value chain: upstream supply chain (such as agriculture), direct operations (such as bottling plants) and product use (such as personal care products).

And the agricultural sector is ripe for improvements in water efficiency – the move to "precision agriculture." As previously discussed the agriculture uses about 70 percent of global water use and in some countries this is much higher. Moreover, water use in the agricultural sector is not efficient due to practices such as flood irrigation and center pivot irrigation.[21]

However, this is changing. The majority of irrigation in the global agricultural sector is from flood irrigation (Figure 5.5). As illustrated in Figure 5.6, flood irrigation is also very inefficient in how it delivers water for agriculture.

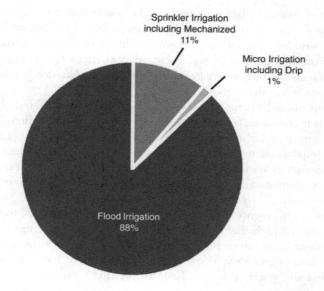

Figure 5.5 Global irrigated farmland by type
Source: adapted from Citi Water Sector Handbook, May 24, 2011

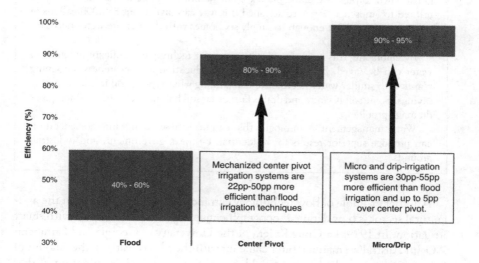

Figure 5.6 Irrigation efficiency by type
Source: adapted from Citi Sector Handbook, May 24, 2011

The move to more efficient water is on. Let's start with an early entry into the movement to "precision agriculture": Netafim.[22] Simcha Blass, an Israeli water engineer, who "discovered" drip irrigation almost by chance, started Netafim. Simcha noticed that a slow and balanced drip effect led to improved plant growth, which led him to create a drip-based tube that slowly released water where it was most effective.

He then piloted his new discovery by launching it at Kibbutz Hatzerim in Israel. A manufacturing facility was established in 1965. Since then, Netafim has been innovating in smart irrigation systems and their technologies have spread globally. Improvements in drip and micro-irrigation technologies have been developed such as systems that adapt to the variations of inlet pressure, clean themselves automatically, maintaining uniform flow rate regardless of water quality and pressure.

It is also worth mentioning that this is part of the movement towards "more crop per drop" which includes the use of drought resistant crops coupled with more efficient irrigation technologies. Companies such as Syngenta,[23] Pioneer,[24] Arcadia,[25] and others have or are developing crops that have greater tolerance of droughts and water scarcity – basically, getting more crops from less water (Box 5.1).

Box 5.1 Xeriscaping

Xeriscaping is a method of landscaping that conserves water. It means instead of a grassy lawn, drought-tolerant plants, stone walkways, mulch, and dirt would be utilized to landscape. An acre designed this way can save about 850,000 gallons of water annually; about enough to supply six homes with their entire water needs for a year.[26]

By embracing this type of water management technique significant amounts of water can be saved. It largely boils down to education and awareness. Grouping plants with similar water needs or understanding what type of soil to use can bring savings. Some soil is sandy and drains fast; other soil has more clay in it, which poses drainage problems.

Water management technologies that can take these savings into consideration can produce superior results for residential, business, government, and industrial grounds.

Another example of how technology can increase water efficiency in the agricultural sector is how detailed topographic mapping can be integrated with smart irrigation. In 1999, Dr Craig Kvien, of the University of Georgia and FarmScan AG, an Australian manufacturer came up with the idea of varying the amount of irrigation water applied across a field, in light of a detailed examination of that field's characteristics.[27] This new approach is called variable-rate irrigation (VRI).

This technology was also born from an observation. While observing crop cultivation in Georgia, Dr Kvien noticed that there was a tendency for the basin's crops to grow in circular patches. This is caused by the way they are watered, for the predominant irrigation systems use sprinkler heads attached to hoses that dangle from wheeled trusses which move in a circular pattern around a central tower (center pivot irrigation).

VRI is now used by more than 80 farms in Georgia, and farmers in Australia, Germany, New Zealand, South Africa, and Spain are taking an interest. Two other firms – Zimmatic and Valley Irrigation, both based in Nebraska – have joined FarmScan AG in selling the kit. The VRI system requires the farmer to produce a map of his land with a resolution of less than a meter, to determine its topography – particularly any low-lying areas where water might pool and higher spots that are prone to run-off. Fallow areas, uncropped parts, watercourses, dirt tracks and wetlands also need to be fed into the system. For further precision, a farmer can use soil-moisture probes to let him know how much water each bit of a field is using, since denser, clay-based soil requires less irrigation than looser, sandy soil.

The farmer, or his agent, uploads all this information into software written by one of the companies involved. This software uses the GPS satellite whose signals provide precise location information to monitor the position of each sprinkler head as it turns on the pivot. The software then works out, on a real time basis, how much water should be released from each irrigation head at any given moment.

The bottom line is that the payoff is an average 15 percent reduction in water consumption, a reduction in fertilizer use because of reduced run-off, and savings of US$40 to 110 per hectare. This is compared with an investment of $5,000 to $30,000 for the VRI system.

Not a bad return when you consider the increasing competition for water and resultant potential limits to availability.

So what about domestic water use and improved efficiencies?

We are seeing increased water efficiencies in domestic water fixtures. For example, the USEPA launched WaterSense.[28] The USEPA WaterSense program is a partnership program designed to offer consumers a simple way to use less water with water-efficient products, new homes, and services. "WaterSense brings together a variety of stakeholders to:

- promote the value of water efficiency;
- provide consumers with easy ways to save water, as both a label for products and an information resource to help people use water more efficiently;
- encourage innovation in manufacturing; and
- decrease water use and reduce strain on water resources and infrastructure.

The goal of the program is to help consumers make smart water choices that save money and without compromising performance. Products and services that have earned the WaterSense label must be at least 20 percent more efficient in water use without sacrificing performance.

According to the USEPA, "if one in every 10 homes in the United States were to install WaterSense labeled faucets or faucet accessories in their bathrooms, it could save 6 billion gallons of water per year, and more than $50 million in the energy costs to supply, heat, and treat water."[29]

The US is not alone in promoting domestic water use efficiency. A couple of examples from elsewhere:

- UK Water Supply (Water Fitting) Regulations of 1999: This is a national requirement for the design, installation, composition and maintenance of water fixtures and fittings. These regulations intend to protect customers and the environment from poor water quality and the misuse of water supplies.[30]
- Australia's commitment to household water conservation: In addition to mandatory water restrictions in many parts of the country, Australians have been voluntarily conserving water by adopting water saving practices and installing water saving devices such as dual flush toilets. The AU$250 million National Rainwater and GreyWater initiative aims to help people use water wisely in their everyday lives including rebates of up to AU$500 for a household that installs rainwater tanks or gray water systems.[31]

One of the ways in which domestic water conservation can be promoted is that governments and water utilities can impose a volumetric water charge on households. This requires that: households have water meters (ideally smart meters); and that household water bills depend on the amount of water consumed. Water efficiency best practice voluntary standards, legislation and subsidies, will be critical in increasing water efficiency.

There is promise with water metering. Smart water meter monitors have been developed that transmit water use performance to inside your home and some monitors are equipped with leak sensors and send out an alert.

By using a display that shows water use volumes, people use less. It turns out that once you see how much water you use, you tend to try to use less. And when that amount can be calculated and displayed to you, say, via a convenient display in your home it can take on a game-like quality of conservation – not unlike the dashboard display in an electric vehicle or hybrid.

Innovators are counting on this disposition. According to a Pike Research report on the smart water meter market, with "increasing demand for water itself, the aging system infrastructure, and a need among utilities to operate their systems much more efficiently," the smart water meter market looks set to expand.[32]

Existing infrastructure requires maintenance and utility operators will also want to evaluate the intrinsic worth of upgrading to the latest meter technology as well. Other factors that will propel smart water meter shipments include the need to conserve scarce water supplies and the need to reduce high levels of non-revenue water as well as the need to satisfy regulatory requirements. Growth will also come from emerging markets in Asia Pacific and elsewhere as water metering rises along with rising standards of living and the need to manage this valuable resource efficiently:

> The necessity for water utilities to adopt smart metering is as pertinent as ever. Smart meters can cut leakage levels by 20 percent and reduce energy consumption by a third through associated software and infrastructure. European utilities alone will invest at least $7.8 billion in smart water metering by 2020 and the market opportunities for meter manufacturers, installers, data and management organizations is clear.[33]

A smart water system can also track weather reports before turning on landscape water systems. It can also reduce water use by 59 percent and water run off by 71 percent. According to Zane Satterfield and Vipin Bhardwaj, who published a brief on the importance of water metering for the National Environmental Services Center:

> Any viable business must be able to determine how much product it is making and selling and if that product is profitable. Water is a business. And, the best way for a water utility to measure or account for the water produced and then sold is by using water meters.[34]

Satterfield and Bhardwaj point to six reasons why water meters are important:

- They make it possible to charge customers in proportion to the amount of water they use.
- They allow the system to demonstrate accountability.
- They are fair for all customers because they record specific usage.
- They encourage customers to conserve water (especially as compared to flat rates).
- They allow a utility system to monitor the volume of finished water it puts out.
- They aid in the detection of leaks and waterline breaks in the distribution system.[35]

Already, companies are capitalizing on the opportunity. Companies in the water smart metering sector include Neptune, Senus SA, Elster Metering, Master Meter, Badger Meter, and Itron. As of 2010, the US has about 30 percent of automated water meters installed globally.

While the benefits of smart meters are clear, adoption of this technology is a challenge. Here is a perspective from Emily Ashworth, a global information technology executive:

In general, from a financial side, the water bill is the lowest of the standing utility-type bills in a household. So the percentage increases that it takes to make a significant technology investment are sizeable and very noticeable by the consumer because of the small base. By comparison, if you have an electric bill, with a much larger base, the percentage increase to make the same technology investment is much smaller to the consumer. This is compounded by the rapid speed with which technology investments depreciate and frankly last before replacement is required. So the financial picture is a challenging one in the water space to begin with.

Another challenge is the relative impact that "the hot" technologies can have in the water space. The lack of applicability leads to complacency in looking for other opportunities. For example, smart metering technologies are a clear focus area for electric utilities. To start the discussion, let's separate out customers who have yard watering needs for a moment. For customers that are general consumers of water (hygiene, cooking, cleaning), general conservation behaviors don't result in significant savings, thus both diminishing the interest in real time data (that which a smart meter can provide) and maintaining the discipline about conservation efforts. The activities that the average consumer can do (turn off the water when brushing teeth, take shorter showers, etc.) don't result in dramatic savings. This is then both met with a general lack of interest by the consumers and a willingness to provide rates to cover the investment by the public utility commissions. If you consider the water consumers with yards, tiered rates have generally proved that consumers don't water their yards less; they are willing to just pay more. So, the real impetus for technology needs to be motivated by sources other than consumer conservation or bill reduction. Smart metering could help with leak detection, the ability to do remote turn on and turn offs, determine when consumers are and are not home for service calls or collection activities, etc. Much of this thinking gets lost in the earlier conversation.

Then there is the engineering mindset – water utilities have been around longer than modern PCs and software systems. All water operations were "owned" organizationally by engineering or operations departments. As technology evolved to serve these areas, these same departments too owned them. Simultaneously, the evolution of the PC and WANs and LANs and email evolved in the corporate world and these items were "owned" and maintained by what would become IT. However, over time the technology to support water operations (SCADA systems, digital meters, hand held devices, AMR, AMI, maintenance systems, GIS, etc.) has evolved and developed, however in many circumstances it is still maintained by people of engineering and operations disciplines who are not IT professionals and have not kept pace with changes in technology or the benefits. So the green screen lingers on.

The other challenge is that utilities have had the good fortune for years of largely recovering capital investments in rates and the rate structures providing coverage of operating expenses, so the search for new revenue sources hasn't been a driving factor in technology innovation. Opportunities to generate revenue by providing functions to other organizations (SCADA monitoring for example) as a revenue source based on a technology based core-competency would be a perfect example, and a need both of other utilities and commercial industries, but there hasn't been the compelling event in water to drive such channel development. If and when a deregulation model comes to water, then there might be more movement in this space.[36]

A 2012 report by Sensus (www.sensus.com) titled, "Water 20/20 Bridging Smart Water Networks into Focus" expands on insight into the challenges with adoption of smart metering.[37] The Sensus report was based upon in-depth interviews and surveys with 182 global water utilities and analyzed utility operations and budgets. The analysis by Sensus identified up to $12.5 billion in annual savings. From the Sensus report:

- *Improved leakage and pressure management:* Globally, about one-third of utilities report a loss of more than 40 percent of clean water due to leaks. It is estimated that by reducing leaks by 5 percent, coupled with up to a 10 percent reduction in pipe bursts, can save utilities up to $4.6 billion annually. By reducing the amount of water leaked, smart water networks can reduce the amount of money wasted on producing and/ or purchasing water, consuming energy required to pump water and treating water for distribution. Intelligent solutions can make a difference. The use of different types of smart sensors to gather data and apply advanced analytics, such as pattern detection, could provide real-time information on the location of a leak in the network.
- *Strategic prioritization and allocation of capital expenditures:* Employing dynamic asset management tools can result in a 15 percent savings on capital expenditures by strategically directing investment. Such tools can save up to $5.2 billion annually. To close the gap between the capital spending required and the amount of financing available, utilities need access to information to better understand the evolving status of their network assets, including pipes.
- *Streamlined network operations and maintenance:* By implementing smarter technology that provides the critical data, via remote operations, utilities can save up to $2.1 billion annually, or up to 20 percent savings in labor and vehicle efficiency and productivity. A smart water network solution can help streamline network operations and maintenance by automating tasks associated with routine maintenance and operation of the water distribution system.
- *Streamlined water quality monitoring:* Smart water networks can save up to $600 million annually, or 70 percent of quality monitoring costs, and far more in avoided catastrophe. A smart water network solution for water

quality monitoring would enable utilities to automatically sample and test for water quality and intervene quickly to mitigate potential issues. By implementing such a system, utilities can incur lower costs from labor and equipment needed to gather samples, as well as a reduction in the amount and cost of chemicals used to ensure regulatory quality standards.[38]

While the savings are impressive, the key aspects of the Sensus report reside in the challenges in technology adoption. To summarize:

- *Lack of a strong business case:* Sixty-five percent of survey respondents frequently cited unfavorable economics or the lack of a solid business case as key barriers to adoption of smart water networks.
- *Lack of funding even if there is a business case:* Sensus proposes possible solutions to lower the barrier to entry include risk-sharing contracts to lower upfront investment required and third-party suppliers who manage and analyze the data.
- *Lack of political and regulatory support:* Input from utilities that what is required is regulatory support and incentives is critical to kick-starting smart water management.
- *Lack of a clear, user-friendly integrated technology solution:* Again from the utility perspective, fragmented product and services offerings from various vendors make it difficult for utilities to integrate a common business plan across their disparate operating divisions.

The challenges in adopting smart water metering are significant. However, unless we can measure water use and inefficiencies and couple this with fair water pricing we will be limited in thinking only about increasing water supply as a solution to increasing demands for water.

The USEPA sees both the need to increasing water supply and demand and the interplay between the two. According to the USEPA:

improving water efficiency reduces operating costs (e.g. pumping and treatment) and reduces the need to develop new supplies and expand our water infrastructure. It also reduces withdrawals from limited freshwater supplies, leaving more water for future use and improving the ambient water quality and aquatic habitat.

More and more utilities are using water efficiency and consumer conservation programs to increase the sustainability of their supplies. Case studies demonstrate substantial opportunities to improve efficiency through supply-side practices, such as accurate meter reading and leak detection and repair programs, as well as through demand-side strategies, such as conservation-based water rates and public education programs.

Accounting for water is an essential step toward ensuring that a water utility is sustainable. Metering helps to identify losses due to leakage and also provides the foundation on which to build an equitable rate structure to ensure adequate revenue to operate the system.[39]

Water metering not only provides the benefit of promoting water efficiency it also provides real time data on water system losses through leakage:

> National studies indicate that, on average, 14 percent of the water treated by water systems is lost to leaks. Some water systems have reported water losses exceeding 60 percent. Accounting for water and minimizing water loss are critical functions for any water utility that wants to be sustainable.[40]

An example of how utilities are beginning to develop a better understanding of water leakage is the work of the American Water Works Association.[41] The AWWA represents suppliers: 57,000 members in 43 sections, and in 100 countries outside of North America. These members provide about 85 percent of the North American population with safe drinking water.

According to the AWWA, many water utilities typically do not tabulate all of such data as the volume of water supplied, customer consumption, distribution system attributes and quantities of losses. Two broad types of loss occur in drinking water utilities:

- *Apparent losses* are due to customer meter inaccuracies, systematic data handling errors in customer billing systems and unauthorized consumption. "In other words, this is water that is consumed but is not properly measured, accounted or paid for. These losses cost utilities revenue and distort data on customer consumption patterns," , the AWWA explains.[42]
- *Real losses* are those types of losses that we can see: leakage and overflows. "These losses inflate the water utility's production costs and stress water resources since they represent water that is extracted and treated, yet never reaches beneficial use," according to the AWWA. The website continues with research that "has found that past practices of defining and calculating "unaccounted-for" and the "unaccounted-for percentage" varied so widely in utilities around the world that these terms had no consistent meaning. Additionally, the unaccounted-for percentage indicator is mathematically misleading and reveals nothing about water volumes and costs, the two most important factors in water efficiency assessments.[43]

To this point, the AWWA developed a water auditing system and believes that water supplied by water utilities can be accounted for, via metering or estimation, and therefore there is no such thing as water that is unaccounted for.

The challenging part, of course, is identifying and monitoring water from source to tap. And this is where businesses can prosper. There is a significant need for tracking services. It reduces costs and increases income. There isn't much of a better selling point for a utility, especially if the value-added completely recoups the costs of the accounting service, which should be a logical business case considering the billons of dollars we spend per year on water alone around the world.

Moreover, tracking should be mandated to begin with. Providing quality drinking water is a critical service that generates revenues for water utilities to sustain their operations. But these revenues rely upon efficient systems of customer metering, meter reading, billing and enforcement that prevent consumption data error – and revenue loss – from occurring. Water utility managers can address these losses by first assessing their policies and mapping the workings of the customer billing system, the AWWA advises.

AWWA also asks:

> Do policy loopholes exist that allow water to be taken without the knowledge and authorization of the utility? Do all customers exist with an account in the customer billing system (whether or not the system is metered)? Are customer meters replaced before they lose accuracy from wear? Is meter reading accurate and complete? Taking time to compile a flow chart of current activities and comparing them to the expectations for revenue recovery usually uncovers a number of shortcomings that can be corrected to recover lost water and revenue. Often these corrections require inexpensive procedural or programming changes that can recoup considerable uncaptured revenue and successfully launch a water loss control program.
>
> Rapidly advancing technology in Automatic Meter Reading (AMR) Systems and Automated Metering Infrastructure (AMI) offers outstanding capabilities to water utilities to improve their efficiency in capturing customer consumption data, identifying wasteful usage and leakage, and other enhancements to improve revenue capture and manage water and revenue losses.[44]

What's tragic and intriguing at the same time is the fact that many water utilities respond to leaks only after a leak has been reported. The AWWA refers to this as "reactive leakage response." And such a leak will likely never be reliably contained. To effectively control leakage, "proactive" leakage management programs need to be implemented. These include the means to identify hidden leaks, optimize repair functions, manage excessive water pressure levels, and upgrade piping infrastructure before its useful life ends.[45]

Many effective technologies have been developed in recent years to quantify and better analyze leakage amounts. These technologies include leak noise "correlators" and loggers, as well as pressure management systems to reduce leakage. "Many effective strategies now exist to allow water utilities to identify, measure, reduce or eliminate leaks in a manner that is consistent with their cost of doing business," the AWWA says.[46]

The importance of increased water efficiency and reduced water leakage can have a substantial impact in addressing global water needs. For example, one of the targeted solutions put forth by the World Water Forum commission is to raise water use efficiency by 15–25 percent by 2020 to meet increased demand and ensure not only water security but also food security. When a pipe breaks or leaks occur, it doesn't just mean that toilet flushing is interrupted or the taps stop working, it can mean that irrigation to crops stops. And that means food sources are threatened.[47]

The recognition that smart metering can increase water efficiency and reduce leakage (non-revenue water) is creating technology opportunities such as Sensus and other global companies. For example, the UK Company Sensornet utilizes fiber optics to detect leaks based on minute temperature changes.[48] The company explains:

> When a leak occurs from a pipeline, the fluid or gas will contain a tempera-
> ture signature which will differ from the surrounding environment. In some
> cases, the temperature changes can be very small (in the case of water leak-
> age in dams) and in other cases the temperature difference can be substantial
> (in the case of liquid natural gas, 120–160°C, or for ethylene 110°C). By
> detecting the temperature change of the surroundings, the distributed
> temperature sensor can not only detect the presence of a leak, but can also
> pinpoint the location of the leak to within 1 to 5 meters.[49]

Sensornet and other companies that offer related technologies of course work on major pipelines in remote places but water leak detection technologies can also be used for domestic water use. For example, TaKaDu is another example of the opportunity in leak detection technology.[50] TaKaDu was founded with the vision of addressing the challenges of water networks worldwide using advanced technology.

TaKaDu has developed a novel approach for understanding the "behavior" of the water network and taking immediate action. TaKaDu's network monitoring service takes data from existing sources and analyzes it in real-time. It automatically locates and classifies water network events as early as possible. Increasing network efficiency and reducing water loss also saves energy. The water sector is typically a country's largest energy consumer. Identifying problems earlier and acting promptly can yield double-digit reductions in energy consumption.

TaKaDu can use multiple sources of raw data: online sensor data (flow and pressure), network operations data, network structure (GIS) and external data (e.g. weather, holidays, scheduled fixes). The monitoring system uses multiple anchors for data analytics: complex periodic patterns, cross-site and spatial correlation, graph structure, event behavior over time and space, pressure-flow relations and many other statistical anchors. TaKaDu's water network monitoring service is available over the web. Water utilities worldwide receive on-line alerts about water network events and insightful information about the state of the network.

Once leaks are detected, how can they be repaired? This, of course, represents another market opportunity.

Here are a couple of examples of innovation in repairing leaks from 3M and Insituform. Both 3M and Insituform have developed technologies to repair water and wastewater pipelines. In-situ repairs represent an economic option when compared to alternative water infrastructure rehabilitation methods. 3M's "line of products and unique trenchless technology application processes are designed to restore pipe width and increase flow rates – minimizing loss and energy expenditure throughout the system."[51]

Insituform Technologies "is a leading worldwide provider of cured-in place pipe (CIPP) and other technologies and services for the rehabilitation of pipeline systems. Insituform's businesses consist of sewer, water and industrial trenchless pipeline rehabilitation and protection."[52]

Expect to see innovation in cost-effective technologies to reduce and eliminate "non-revenue" water and wastewater pipe leakage. Perhaps not very exotic compared to technologies such as desalination, but no less critical.

Tackling water efficiency and water leakage losses represent a real business opportunity. Billions of dollars worth of municipal and private projects are coming down the pike – and they'll be seeking water efficiency and leakage detection/repair solutions.

Water reuse, recycling, and desalination

This is where water tech gets interesting: increase reuse, increase recycling and tap into other sources of water (brackish and saline) to "create" freshwater.

Water is the *ultimate renewable resource*; for the most part you can continue to reuse and recycle the water indefinitely. We are moving to a paradigm where water efficiency coupled with aggressive reuse and recycling will be the norm and not the exception.

What are the big moves in water reuse and recycling? What role does desalination play in addressing water scarcity? Let's tackle these questions now.

According to the European Commission AQUAREC report,[53] the benefits of water reuse and recycling are significant in view of increasing water scarcity. Treated wastewater has been:

> an important means of augmenting river flows in many countries and the subsequent use of such water for a range of purposes constitutes indirect reuse of wastewater, it is becoming increasingly attractive to use reclaimed or treated wastewater more directly.
>
> Treated wastewater may be used as an alternative source of water for irrigation in agriculture. Agriculture represents up to 60 percent of the global water demand while the requirements arising from increasing urbanization such as watering urban recreational landscapes and sports facilities also creates a high demand. Water scarcity in Mediterranean countries historically led these countries to use appropriately treated wastewater in agriculture.

Treated wastewater may also be used as an alternative source of water for irrigation of golf courses and other green spaces, including those used for recreation in which individuals may come into contact with the ground. It can be used to supplement artificially created recreational waters and for reclamation and maintenance of wetlands for which there can be a significant ecological benefit and a subsequent sense of benefit to the community.

Finally, an additional use may be the direct supplementation of drinking water resources through groundwater infiltration and by adding it to surface water. There are several cities in northern Europe that rely on indirect potable reuse for 70 percent of their potable resource during dry summer conditions. It is even technically possible for it to be used as a direct drinking water source, although acceptability to the public may not yet be achievable.[54]

Water treatment is essentially a process of solids removal and chemical treatment stages. Treatment involves the removal of inorganic and organic compounds with several of these compounds requiring special treatment.

Water treatment technologies can typically be categorized into primary, secondary, disinfection, and advanced technologies. A brief overview of these categories of treatment technologies is provided below.

- *Primary treatment*. As the name implies, this is where it all begins. This phase involves the screening of waste materials and separation of solids and liquids.
- *Secondary treatment*. This step is the biological treatment phase and involves membrane bioreactors (MBR). Oxygen is added to increase the growth of microorganisms to consume dissolved organic material in wastewater. If the concentrations of organic material are high then this phase may use an "activated sludge" treatment process. With activated sludge the microorganisms are at high concentrations to ensure digestion of the organic material. Another component of secondary treatment is where the effluent is disinfected with chlorine prior to discharge to surface water or ocean. Increasingly this treated water is now being reused.
- *Disinfection – chlorine and ultraviolet (UV) light*. Historically (since the 1900s), disinfection was achieved with chlorine to kill pathogenic bacteria and eliminate (or reduce) odors. Alternatives to using chlorine are ozone and UV technologies. UV technologies are a rapidly growing alternative to chlorine and ozone.
- *Advanced (tertiary) treatment*. This phase involves the reclamation of wastewater for alternative uses such as irrigation. Tertiary treatment involves the use of sand filters to remove any remaining solids and an additional disinfection process. Tertiary treatment can consist of biological steps to remove nitrogen and phosphorous, chemical separation, carbon adsorption, distillation, deionization and filtration process such as ultrafiltration, microfiltration, and reverse osmosis technologies. Tertiary treated water can

be reused for industrial, agricultural, recreational or drinking water (remember Singapore?).

A simplified view of how we move water around and the water treatment ecosystem is illustrated in Figure 5.7. Although we will discuss this in greater detail in Chapter 6 (the energy–water nexus), remember that every step illustrated in Figure 5.7 requires energy. The more efficient the steps (or eliminating the steps), the lower the energy use.

The steps in the extraction, treatment, use and discharge are:

- *supply* – surface or groundwater;
- *treatment* – filtration and disinfection;
- *use* – residential, commercial and industrial;
- *end* – handling (and treatment) of water prior to discharge; and
- *discharge* – returns to surface water, groundwater, ocean or evaporation.

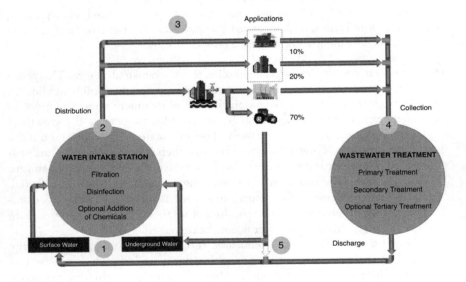

Figure 5.7 The basic water life cycle[55]

One of the big moves in the water industry is towards POU treatment. The reasons are simple. Currently, treating water at centralized locations to primary drinking water standards is energy intensive. The alternative would be to collect and treat water as needed for distribution to the point of use where it would be treated according to the use requirements and then re-used as needed.

There is also a movement by consumers to move towards a POU treatment. The options are to treat the water as it enters the home for softening, for example. Options for home treatment include the Pentair whole home water treatment system (SmartWater) or the Culligan drinking water system. Systems range from those mounted on faucets to a larger scale for commercial use in restaurants and hotels.

Industrial treatment of water includes a wide range of applications in the mining, food and beverage, paper and pulp, refining, power generation and commercial HVAC (heating, ventilation and air conditioning) systems. These treatment options use chemicals to prevent corrosion, scaling and microbial growth. The water is then treated prior to discharge.

Here is a brief overview of some of the trends in water recycling/reuse and desalination:

- *Desalination.* Increased demand for freshwater is driving a renewed interest in desalination technologies coupled with progress in reducing the energy footprint of desalination.[56] The obvious appeal of desalination is to tap into saline and brackish water sources to meet the ever-increasing demands of urbanization, industrialization and energy requirements (shale gas fracking, for example) without the threat of impacts from droughts.

  Innovation in pre-treatment systems coupled with increased membrane and pump efficiencies is reducing the costs of desalination from about $1.50 per cubic meter to about $0.70 per cubic meter.[57] In addition, technology advances in reverse osmosis have contributed to increased efficiencies and, as a result, increased interest in the desalination technology. Growth in adopting desalination technology has grown by about 9 percent per year since 2005 and is expected to grow by about at double-digit rates through 2015.

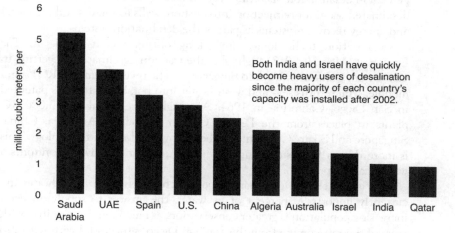

Both India and Israel have quickly become heavy users of desalination since the majority of each country's capacity was installed after 2002.

Figure 5.8 Top ten countries by installed desalination capacity since 2003[58]

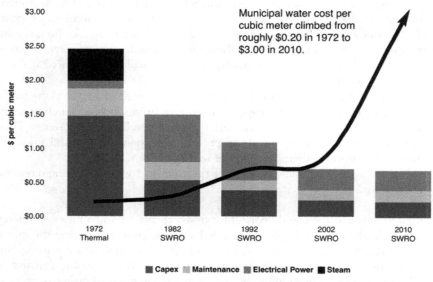

Figure 5.9 Cost of thermal versus reverse osmosis desalination from municipal water from 1972 to 2010[59]

Trends with desalination technologies include; a move to "zero-discharge" systems to address the discharge of salt byproducts, co-location of desalination plants and power pants (as energy can represent about 50 percent of desalination operating costs), offshore desalination systems (using desalinated water for reinjection into offshore wells to increase oil recovery) and energy recovery systems as part of the desalination systems.

Desalination technologies are taking off. A recent article in *The Economist* (May 31, 2012) highlights the traction desalination is getting in places from California, USA to Singapore.[60] The reverse osmosis (RO) spiral module, which turns sea- and waste-water into potable water, was patented in San Diego, California in 1964. Although there are about 13,000 RO plants in places from the Persian Gulf and Israel to Australia, China, Singapore and Spain, California has been slow to adopt the technology. This is more than a little surprising since the American southwest confronts a water scarcity problem.

This is now changing. In San Diego, which has brackish groundwater and currently imports 90 percent of its water, the answer to supply an ever-increasing population is greater conservation, as much reuse as possible, with most of the rest coming from the sea. San Diego is successful with regard to conservation – today the city uses less water with a larger population than it did in 1989, the year water consumption peaked. However, water recycling has been more challenging.

This is where desalination comes in. Poseidon Resources is now close to building the biggest desalination plant in America behind a power station by the beach in Carlsbad, California. The power plant uses 304 million gallons of seawater a day for cooling, so Poseidon plans to divert 104 million gallons a day through its osmotic membranes. Of the 104 million gallons per day, 50 million gallons of freshwater will be created. Once running at full scale by 2015, the plant could produce 10 percent of the region's water. And there are plans for more desalination plants which have far reaching benefits – inland Californians, Arizonans, Nevadans and others would need to take much less water from the water-starved Colorado River.[61]

- *Water reuse/recycling.* One of the main aspects of a move towards 21st Century paradigm for water is water reuse. Water can no longer be used as a commodity that is used once and then discarded. Currently, treated water is typically discharged to surface waters, the land, reinjected into aquifers or the ocean. This really just doesn't make any sense. As a result, expect to see water reused for irrigation, heating, cooling and other non-potable applications. In particular, reused water has applications in the mining and paper and pulp sectors.

- *Membranes replacing chemicals in water treatment.* The big driver here is an advance in material science and nanotechnology. Membrane technologies are becoming more efficient in filtering out complex contaminants and compounds such as runoff from agricultural lands and pharmaceutical byproducts in residential wastewater. Growth in these membrane technology applications is projected to increase from $1.5 billion in 2009 to about $2.8 billion by 2020, which is a 6 percent ten-year compound annual growth rate (*Citi Water Sector Handbook*, May 24, 2011). The range of filtration/membrane technologies is illustrated in Figure 5.10.

- *Forward osmosis.* Forward osmosis is similar to reverse osmosis in that it uses a semi-permeable membrane to separate water from dissolved solutes. The difference is that forward osmosis uses naturally occurring osmotic pressure between salt water and freshwater to drive the separation. In contrast, reverse osmosis uses external hydraulic pressure from pumps, which requires a significant amount of energy. The advantages of forward osmosis are that there is little fouling of the membrane, increased separation of a wider range of contaminants and its relatively low energy footprint. The forward osmosis technology can use the same membranes as RO and the membranes can be reused.

- *UV replacing chorine as a disinfectant.* In the US since the late 1990s the USEPA has encouraged the use of disinfectants other than chlorine. The UV light does not kill the parasitic disease but instead renders them unable to reproduce and infect humans. The UV also has no residual benefit – after the UV is applied there are no ongoing disinfection benefits. In contrast chlorine will continue to disinfect the water for days following application.

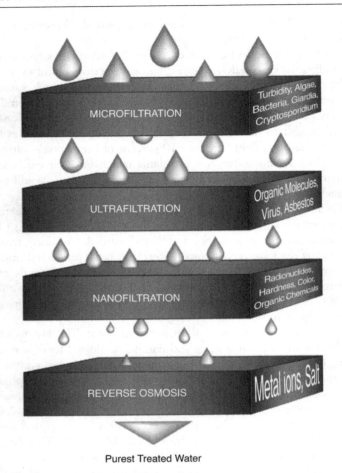

Figure 5.10 Profiles of water filtration/membrane applications
Source: adapted from Citi Water Sector Handbook, May 24, 2011

- *Point of use treatment.* As previously discussed there is an increasing move to treating water at the point of use. There are really a couple of drivers for a movement to point of use treatment – the inefficiency in treating water at a centralized location and then transporting the water trough leaking pipes coupled with a concern by some consumers about bottled water. The increase in POU treatment may actually be more significant in emerging markets where centralized systems have yet to be installed.

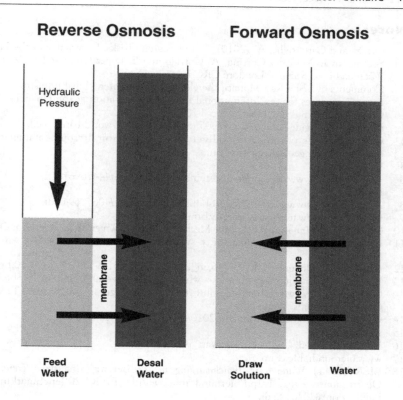

Reverse Osmosis   **Forward Osmosis**

Figure 5.11 Use of natural osmotic pressure for forward osmosis
Source: adapted from Citi Water Sector Handbook, May 24, 2011

It is also worth noting the opportunity in water testing. The testing of water quality is moving from periodic testing to inline and continuous testing of water quality for centralized and decentralized POU systems. This would only make sense as public concerns increase on water quality and the drive towards water reuse and recycling – we all want to know the quality of the water we ingest or use in our industrial and commercial applications.

So where are water reuse, recycling and desalination headed? For a range of reasons we are moving towards; increasing our supply of freshwater through desalination (RO and FO), moving away from chemical treatment (UV and membrane technologies), moving away (to the extent possible) from large decentralized water treatment systems, and no longer handling water as a disposable commodity (reusing and recycling water as much as possible).

Watch these trends and the business opportunities that they foster.

Notes

1   Orr, S. and Cartwright, A. (2010) "Water Scarcity Risks: Experience of the Private Sector," in L. Martinez-Cortina, A. Garrido and E. Lopez-Gunn (eds), *Re-thinking Water and Food Security*, London: CRC Press.
2   Comments by Usha Rao-Monari, the global head of water, global infrastructure and natural resources for the International Finance Corporation at the Financial Times Conference on Sustainability, 2 April 2012.
3   http://bluelivingideas.com/2010/05/05/raise-price-water-world-bank-oecd.
4   www.oecd.org/environment/biodiversitywaterandnaturalresourcemanagement/thewaterchallengeoecdsresponse.htm.
5   Ibid.
6   Adapted from www.oecd.org/tad/sustainable-agriculture/44476961.pdf.
7   Ibid.
8   Adapted from www.oecd.org/tad/sustainable-agriculture/40678694.pdf.
9   www.fao.org/nr/water/issues/scarcity.html.
10 OECD (2010) *Innovative Financing Mechanisms for the Water Sector*, Geneva: OECD.
11 www.huffingtonpost.com/robert-stavins/misconceptions-about-wate_b_175281.html.
12 www.oecd.org/document/30/0,3746,en_2649_34285_45799583_1_1_1_1,00.html.
13 Ibid.; Stavins is quoting from the white paper he co-authored with Sheila M. Olmstead, www.hks.harvard.edu/fs/rstavins/Monographs_&_Reports/Pioneer_Olmstead_Stavins_Water.pdf.
14 www.guardian.co.uk/environment/2010/apr/27/water-price-rise.
15 www.rmi.org.
16 http://en.wikipedia.org/wiki/Negawatt_power.
17 www.bieroundtable.com.
18 BIER (2011) *Water Use Benchmarking in the Beverage Industry: Trends and Observations, 2011*, http://bieroundtable.com/files/BIER%20Benchmarking%20Publication%202012.pdf.
19 Ibid.
20 Ibid.
21 www.ers.usda.gov/publications/arei/ah722/arei2_2/arei2_2irrigationwatermgmt.pdf.
22 www.netafim.com.
23 www.syngenta.com.
24 www.pioneer.com.
25 www.arcadiabio.com.
26 http://liveearth.org/pt-br/liveearthblog/trimming-timing-and-topping-off-conserving-water-outside.
27 The Economist (2012) "Dribbles and Bits," *The Economist Technology Quarterly*, 2 June.
28 www.epa.gov/watersense.
29 Ibid.
30 www.thameswater.co.uk/cps/rde/xchg/prod/hs.xsl/896.htm.
31 http://wealthmanagement.ml.com/Publish/Content/application/pdf/GWMOL/Global-Water-Sector.pdf, page 51.
32 Richard Martin (2012) "Installed Base of Smart Water Meters to Reach Nearly 30 Millions Worldwide by 2017," 5 January, www.pikeresearch.com/newsroom/installed-base-of-smart-water-meters-to-reach-nearly-30-million-worldwide-by-2017.
33 www.prlog.org/11776638-smart-water-meter-market-set-for-growth-over-the-next-five-years.html.
34 www.nesc.wvu.edu/pdf/dw/publications/ontap/2009_tb/water_meters_DWFSOM67.pdf.

35 Ibid.
36 Emily Ashworth, personal correspondence for this book.
37 Sensus (2012) *Water 20/20 Bridging Smart Water Networks into Focus*, London: Sensus, http://sensus.com/web/usca/solutions/smart-water-networks.
38 Ibid.
39 http://water.epa.gov/infrastructure/sustain/wec_wp.cfm.
40 Ibid.
41 www.awwa.org.
42 Ibid.
43 Ibid.
44 Ibid.
45 Ibid.
46 Ibid.
47 www.worldwaterforum6.org/en.
48 www.sensornet.co.uk.
49 Ibid.
50 www.takadu.com.
51 www.3m.com.
52 www.insituform.com.
53 B. Durham, A. N. Angelakis, T. Wintgens, C. Thoeye and L. Sala (2005) *Water Recycling and Reuse: A Water Scarcity Best Practice Solution*, The AQUAREC project (European Commission EVK1-CT-2002-00130), as part of the EUREAU water recycling and reuse working group.
54 Ibid.
55 CitiGroup (2011) *Citi Water Sector Handbook*, May 24.
56 Ibid.
57 Ibid.
58 Adapted from *ibid*.
59 Adapted from *ibid*.
60 The Economist (2012) "Salty and Getting Fresh," *The Economist*, 31 March.
61 Ibid.

Chapter 6

The water, energy, and food nexus

Over the past couple of years there has been considerable discussion of the water, energy, food nexus. I (Sarni) have been fortunate to be involved in the dialog of mapping out solutions to address the water needs for energy and agriculture.

During the fall of 2012, I participated in three workshops in Colorado between energy and water utilities and other stakeholders, focused on coming up with long-term solutions to the needs of the energy and water sectors. These workshops were designed and facilitated by Recharge Colorado.[1] Recharge Colorado was an independent, non-profit organization serving as a self-sustaining resource efficiency asset for regional and national stakeholders formed to promote the advancement of a robust, effective marketplace for energy efficiency, water efficiency, and renewable energy projects in Colorado to the benefit of both consumers and businesses through initial services to include:

(1) A customer platform to find resource efficiency and renewable energy incentives, contractors and information with enhanced custom branding and website integration capabilities for partners, (2) Streamlined rebate administration with greater ability to adapt to partner needs, and (3) Data analysis for designing, targeting, and evaluating resource efficiency programs and policies.[2]

Recharge Colorado and other similar organizations globally are tackling the challenge of competing demands for water. The importance of these efforts was made clearer by the prolonged droughts in the US. At time of writing (spring and summer 2012) the US is in the midst of a prolonged drought, experiencing moderate to exceptional drought conditions. As we discussed, we can experience water scarcity without drought; the US drought gives us more than a glimpse into what happens when water is scarce.

In the US, there is increasing concern as to how water scarcity (the drought) will continue to impact food prices and energy production. Energy production from nuclear power plants, biofuel (or "agrifuel") production (as mandated by the USEPA), and the need for freshwater for shale gas fracking are in competition for scarce water supplies.

A few facts from the World Economic Forum on the increasing demand for energy, water and food:[3]

- Population is expected to increase from about 7 billion to 8 billion in the next two decades.
- Economic growth will be about 6 percent in developing economies and about 2.7 percent in developed economies.
- Urbanization will continue to increase; greater than 50 percent of the world's population now lives in an urban environment.

These three powerful drivers are resulting in demands on agriculture, energy and water needs.

- *Agriculture* – To meet growing demand, the agricultural sectors will need to increase production by 70 to 100 percent. This coupled with changing diets (a shift to increased consumption of meats) will result in increased water use.
- *Energy* – Increased demand for energy. The global economy will require about 40 percent more water by 2030 than today.
- *Water* – Increased rise in economic wealth increases water use – from 1990 to 2000 the world's population grew by a factor of 4 but water use increased by a factor of 9.

This tension between energy, water, and food is resulting in some companies taking a position that there should be "no food for fuel" – we can't grow food to burn for transportation. The chairman of Nestlé, Peter Brabeck-Letmathe, has clearly articulated this position.[4] This debate is just warming up as we attempt to solve the challenge of feeding and providing energy and water for an increasing global population.

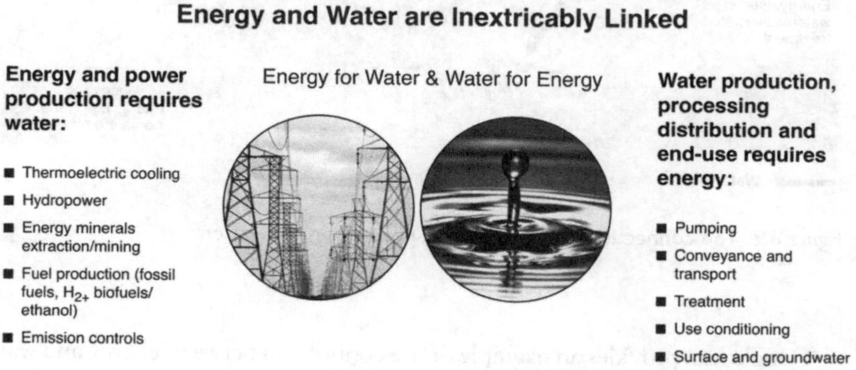

Energy and Water are Inextricably Linked

Energy and power production requires water:

- Thermoelectric cooling
- Hydropower
- Energy minerals extraction/mining
- Fuel production (fossil fuels, H_2, biofuels/ethanol)
- Emission controls

Energy for Water & Water for Energy

Water production, processing distribution and end-use requires energy:

- Pumping
- Conveyance and transport
- Treatment
- Use conditioning
- Surface and groundwater

Figure 6.1 The energy–water nexus[5]

Those technologies that can enable economic growth and access to water by addressing the competition for water from the energy, industrial, agricultural and domestic sectors should have a promising future.

Let's start with the basics of energy and water.

Whether you are a domestic, agricultural or industrial user of energy, saving energy saves water.

It takes water to cool generators that power the electrical grid. It takes water to cool nuclear power plants, and for solar panels to function. Hydroelectric power, of course, would not exist without water. And so on.

As a result, saving water saves energy. Saving energy saves water. Herein lies the opportunity for technology in a big way.

The connections between energy production and water, and between water and energy, are illustrated in Figure 6.2 below.

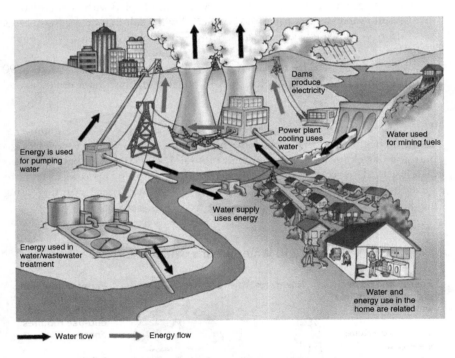

Figure 6.2 The connection between energy production and water[6]

Google Inc. provides an example of the connection between energy and water use. Google relies on data centers to manage all the information it transmits and data centers not only require a tremendous amount of energy to power computers they also need water to cool equipment. The water management system is such a concern to Google that it devised its own water management system. Here's how that works, as stated on their website:

> All of the electricity that goes into a data center ultimately turns into heat, and thus there are fans, pumps, and air conditioning equipment to remove all that heat. Fortunately, there is an energy-efficient way to use water to remove all of that heat. It's called free cooling, and we use it as much as we can.[7]

> On average, two gallons of water is consumed for every kilowatt-hour of electricity produced in the US. By using less electricity to power our computing infrastructure, we also save fresh water. Water used to cool data centers needs to be clean, but not nearly as clean as drinking water. Using recycled water, which we clean just enough for cooling, saves pure water for other uses.[8]

The concept of "free cooling" is important. Chiller-based cooling, which is like utilizing a refrigerator to bring temperatures down, uses energy and water – just like your home refrigerator. In many data centers, chiller-based cooling is responsible for much of the overhead in energy usage. For a company like Google, that overhead expense can be significant. So Google explains:

> We've slashed that percentage by using free cooling techniques to minimize the amount of time that chillers need to run. One example of a free cooling technique is evaporative cooling. Evaporation is a powerful tool. It helps us maintain our body temperature even when ambient temperatures exceed our normal 98.6°F. How? When heat from its immediate surroundings changes water into vapor, the heat dissipates as well, causing a cooling effect.[9]

Evaporating water isn't the only way to "free cool," Google says. For example, its facility in Finland uses seawater to provide chiller-less cooling:

> The site's location on the Baltic Sea and advantageous climate were our main reasons for choosing it. The cooling system we designed pumps cold water from the sea to the facility, transfers heat from our operations to the seawater through a heat exchanger, and returns the water to the gulf. This approach provides all of our needed cooling year round, so we didn't have to install any chillers at all.[10]

By thinking through the water–energy nexus, Google has reduced its energy-related overhead to an average of 16 percent versus the normal overhead of around 90 percent. And this all drops to the bottom line.

And in terms of their water footprint:

Every year our efficient data centers save hundreds of millions of gallons of drinking water simply by consuming less electricity....The idea behind this is simple: instead of wasting clean, potable water, use alternative sources of water and clean it just enough so it can be used for cooling. Cooling water still needs to be processed, but treatment for data center use is much easier than cleaning it for drinking.

The source of waste water can vary. One data center pulls water from an industrial canal while another facility treats city wastewater. We're also investigating capturing rainwater to use for cooling at another facility. While it's not always technically or economically feasible to use recycled water, we're optimistic that we can find sustainable solutions for the majority of our water use. Recycled water isn't widely used in the data center industry today, but we're excited to see others start to adopt sustainable water practices.[11]

These data center practices are increasingly being adopted as reflected in the focus on energy and water by The Green Grid.[12] The Green Grid is a consortium of end-users, policy-makers, technology providers, facility architects, and utility companies to improve the resource efficiency of data centers. It's comprised of more than 175 member companies around the world.

In 2011, The Green Grid proposed a new metric for its members to use. This metric, "water usage effectiveness" (WUE), enables data center operators to quickly assess the water, energy, and carbon sustainability aspects of their data centers, compare the results, and determine if any energy efficiency and/or sustainability improvements need to be made. Over the past few years, The Green Grid has developed a numbers of metrics so data centers can more effectively manage energy on a standardized basis. The metrics are widely adopted within the technology industry.

According to The Green Grid's website, their water metric will address water usage in data centers, "which is emerging as extremely important in the design, location, and operation of data centers in the future."[13]

The Green Grid seeks to unite global industry efforts, create a common set of metrics, and develop technical resources and educational tools to further its goals. Companies such as IBM, Microsoft, Dell, Intel, eBay, and others belong to The Green Grid, so the power of adopting its metrics is felt industry wide.

And that is just the technology industry.

The River Network,[14] an organization dedicated to preserving freshwater supplies, reports that there is a growing awareness that water and energy issues are closely connected. As it reports on its website:

What's not yet widely understood is just how much water we can save by saving energy, and how much energy we can save by saving water. The potential is enormous. With demand for water and energy continuing to grow, addressing the water–energy nexus is an opportunity we can't afford to miss.

The more energy we save, the easier it is to reduce the harmful effects of our greenhouse gas emissions. The more water we save, the easier it is to secure precious freshwater resources and maintain a healthy, climate-resilient environment. Understanding these relationships between water and energy is more important than ever in today's changing world.[15]

Some of the research the River Network has to offer on their page titled The Water–Energy Nexus is startling:

At a minimum, the United States uses the equivalent of 520 billion kilo-watt-hours per year – equivalent to 13 percent of the nation's total electricity use – to pump, heat and treat water. This is double what is generated by all of the nation's hydroelectric dams in an average year and equal to the output of over 150 typical coal-fired power plants! The bad news is that saving energy through water conservation, efficiency and reuse is not currently being utilized as a major strategy for addressing climate change. The good news is that this is one of the largest categories of energy use that we could reduce quickly and significantly, with the added benefit of protecting our water resources from the threat of a changing climate.

We can't eliminate all water-related energy use, of course, but we could reduce a great deal of it in just a few years through water conservation, efficiency and low impact development. We could reduce a great deal more of it over the next few decades if we begin now to replace many uses of treated drinking water with harvested rainwater or treated wastewater.

Water is used in almost every aspect of energy production. In a 2006 report, the Department of Energy estimated that 'in calendar year 2000, thermoelectric power generation [coal, nuclear, natural gas] accounted for 39% of all freshwater withdrawals in the US, roughly equivalent to water withdrawals for irrigated agriculture." The report also states that consumption of water for electrical energy production could more than double by 2030 if current trends persist, equaling the United States' entire domestic water consumption in 1995!

As we shift towards cleaner energy, some alternatives – including biofuels and coal with carbon sequestration – can significantly increase freshwater demands. Luckily, other clean energy technologies such as wind and photo-voltaic solar power use virtually no water, which means clean air and healthy rivers in your community.[16]

What's most compelling about the above is that it just doesn't take water to produce energy; it takes energy to produce water.

In California, for example, it takes about 15 percent of the state's energy just to transport water. According to the California Energy Commission:

Water is first diverted, collected, or extracted from a source. It is then transported to water treatment facilities and distributed to end-users. What happens during end use depends primarily on whether the water is for agricultural or urban use. Wastewater from urban uses is collected, treated, and discharged back to the environment, where it becomes a source for someone else. In general, wastewater from agricultural uses does not get treated (except for holding periods to degrade chemical contaminants) before being discharged directly back to the environment, either as runoff to natural waterways or into groundwater basins... there is a growing trend to recycle some portion of the wastewater stream – recycled water – and redistributing it for non-potable end uses like landscape irrigation or industrial process cooling.[17]

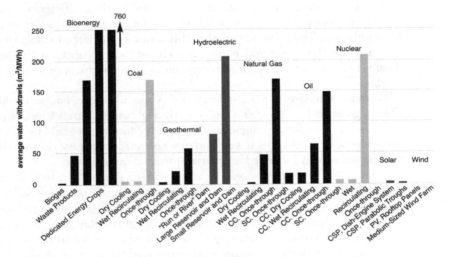

Figure 6.3 Water required for energy production[18]

So while water and energy have a symbiotic relationship in terms of use, they also have the same type of relationship when it comes to savings. Energy-saving initiatives are frequently discussed, but water-saving initiatives less so. But that may change as more and more people, businesses, and governments realize the interconnectivity between water and energy and the price of water aligns with its value.

The growing importance of water is not lost on the energy industry. When the term "energy savings" is tossed about, you can bet the energy industry sits up and takes notice. Increasingly, the energy industry is hearing the word "water" in the same sentence as energy. For example, according to Schlumberger Business Consulting:

> Global population growth and economic expansion will not only create tremendous upward momentum for energy demand, but also drive significant increases in the need for water. Further compounding this challenge is the interdependency and frequent competition between water and energy – large amounts of water are consumed to generate energy, and a vast amount of energy is consumed to extract, process, and deliver clean water. Addressing the water challenge will be a strategic imperative for the oil and gas industry. Water issues are on par with the other major challenges facing the industry, including carbon emissions, the 'big crew change', [sic] limited resource access, geopolitical instability, and increasing technological complexity. Collaboration between industry stakeholders (e.g. operators, service companies, regulators) will be critical to connecting the dots and sustainably addressing the implications of a changing E&P landscape. Water is a limited and critical resource that will influence the way oil and gas companies do business.[19]

Recently, the development of shale gas and associated needs for freshwater have gained much attention. The International Energy Agency (IEA) outlines freshwater demands in the development of shale gas and recommendations in their *World Energy Outlook 2011* report.[20]

Hydraulic fracturing is used to stimulate the flow of gas in shale gas wells, and this technique requires significant volumes of fresh water. The total volume of water injected into shale gas formations ranges from 7,500 to 20,000 cubic meters per well. The IEA recommends the following to address concerns on water and shale gas fracking:

- *Minimize water use.* Improving the efficiency of water use in water-intensive operations through reuse and recycling reduces the burden on local water resources. Given that some of the fracturing fluid injected into wells returns to the surface contaminated by naturally occurring substances that leach from the rocks- minimizing water use can also reduce treatment and disposal needs.
- *Dispose of produced water appropriately.* Because of the sheer volumes involved, enforcing stringent and consistent regulations requiring appropriate treatment before water disposal is the most effective means of minimizing water contamination. Complete disclosure of the chemicals used in the fracturing process would improve the quality of the environmental debate.[21]

The recommendations are being adopted and driving technology innovation not only to increase water reuse and recycling (water for fracking fluids and also treatment of produced water) but also to develop shale gas fracking technologies that do not use fresh water.

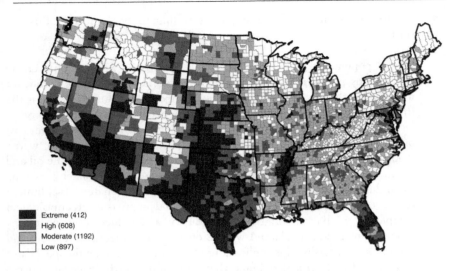

Figure 6.4 Approximate extent of water scarce areas in the US – water supply
sustainability index by county projected through 2050[22]

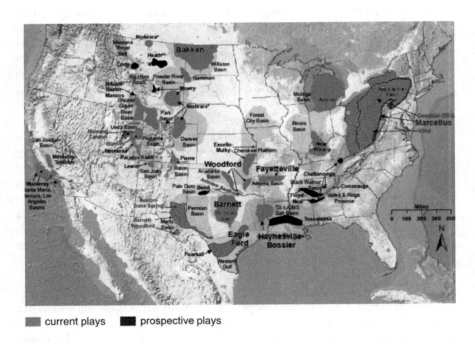

Figure 6.5 Generalized locations for US lower 48 shale gas plays[23]

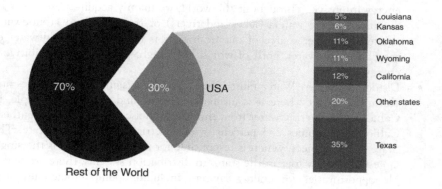

Figure 6.6 Global produced water volumes[24]

What about innovation in the energy–water nexus? A few examples are provided below.

Imagine H2O (www.imagineh2o.org), an organization that incubates technology entrepreneurs and hosts challenges to give start-ups the opportunity for funding, believes the water–energy relationship is worth rebuilding (Pechet is the co-founder of Imagine H2O):

> Delivery and production of clean water currently come at sizable energy costs. The more than 60,000 water systems and 15,000 wastewater systems through the country are huge energy consumers, using 3 percent of annual electricity consumption in the United States. The USEPA estimates that retrofitting one out of every 100 homes with water-efficient technologies could save 100 million kilowatt-hours of electricity per year and avoid adding 80,000 tons of greenhouse gas to the atmosphere. In some parts of the country, you can save more energy by turning off the tap than by turning off the lights.[25]

With energy costs rising, and with fossil fuels constrained by limited supply, Imagine H2O is seeing a bright future for water tech companies that can reduce both water and energy costs. "Provision of water cannot rely on cheap, abundant energy supplies forever," Imagine H2O declares. "Tomorrow's water supply system is an energy-efficient one."[26]

To that end, the organization sees four areas of the water sector prime for innovation:

- *Sourcing Water.* To provide for growing demand, accessing water supplies is of utmost importance, and this first step of acquiring fresh water is very energy intensive. Throughout the world, we mainly acquire fresh water by diverting surface waters (lakes and rivers) or pumping from groundwater aquifers. Another source of our freshwater is in desalinating saltwater or recycling wastewater, both of which have high – and often prohibitive – energy costs.

- *Distributing Water.* Water must often travel over significant distances and elevations to get where it is ultimately treated and used. For example, in California, delivering water from the San Francisco Bay Delta to Southern California consumes 2–3 percent of all electricity used in the state. The State Water Project, which is responsible for this conveyance, is the single largest electricity user in the state. In distribution systems, there are multiple opportunities for energy savings, including pump efficiencies, leak detection, pressure management, and automation.

- *Water Treatment.* Treatment for urban water use has significant energy demands. Further, as source water quality worsens and quality standards tighten, more advanced treatment (such as ozonation and ultraviolet radiation) will require even more energy.

- *Disposal.* Collecting and treating wastewater for proper disposal or reuse calls for a great deal of energy. In fact, almost 30 percent of a wastewater treatment plant's operation and maintenance costs go toward energy provisions. There are many opportunities for innovation in pumping, aeration, and solids handling to save energy in the wastewater treatment process.

All across the water sector there are places where water tech can play a crucial role in direct savings and efficiencies of both water and energy.

Central to saving both energy and water is the notion of "negawatts" and "negadrops." Several US states are embracing the negawatt concept to reduce energy consumption, and attention is being had internationally. The big idea is that those who produce negawatts, or simply conserve energy can earn money by selling or trading the saved energy.

Now let us take this construct to the water market, the "negadrop."

Design Ecology, a blog devoted to landscape architecture and civil engineering, sees a big market for the use of a "negadrop" as a new approach to fixing infrastructure. Negadrops could sidestep regulatory hurdles and political pitfalls by incorporating a new mindset into design from the get-go.

Design Ecology describes it this way in an entry dated 29 March 2009: "The obvious way to create negadrops is by using less water through such technology improvements and dual-flush toilets and waterless urinals. Another approach is to use the water more than once."[27]

It also takes a swipe at "building codes and standards" that disallow the use of recycled water for toilet use. "Simple and available technology would allow us to use rainwater along with bath, shower, and laundry water for toilet flushing cheaply and safely – reducing household water use by 1/3," the blog calculates:

This same approach can be applied to irrigation water – as much as 80 percent of summer residential water use in California.

Not only would this reduce the amount of power required to pump water into the home, but it would also reduce the amount of power required to pump it away from our homes and process the "wastewater." By maximizing both efficiency and on-site water recycling, we produce "negadrops" of water, satisfying secondary water needs without increasing associated energy and water impacts.[28]

Such innovative approaches may be necessary to create the type of conservation that will be needed to address the shrinking supplies of both water and energy sources. The holistic approach, even, may educate people as to the ever flowing and circular relationship between water, waste, and energy. It is those three things that make the world go round. They are necessary parts to our existence and they are manageable. This positions them as commodities. And as such they need to be valued – both in the financial and moral senses of the word.

Taking this further, energy, water, and food security are closely tied.

In an academic paper titled "Scientific Challenges Underpinning the Food-Versus-Fuel Debate," Kenneth Cassman, the director of the Nebraska Center for energy sciences research at the University of Nebraska, notes that, "Because the amount of arable land suitable for intensive crop production is limited, the use of food/feed crops for biofuels is placing tremendous pressures on global food supply and on land and water resources."[29] He finds that although irrigated agriculture produces some 40 percent of global food supply on 18 percent of total cultivated area, "water resources available for irrigation are decreasing due to competition from other economic sectors."[30]

Ensuring an adequate supply of crop commodities for food, livestock feed, biofuels and biobased products without a large expansion of crop area into rainforests, wetlands and grassland savannahs will require massive increases in crop yields on existing farm land. Given these trends, there is an urgent need to accelerate crop yields to rates well above the historical trajectories of the past 40 years, while at the same time protecting soil and water quality and reducing greenhouse-gas emissions.[31]

Given that highly irrigated cereal crops account for nearly 60 percent of all calories in human diets and that the area devoted to cereal crops has decreased by 1.8 million ha per year since 1980, while global expansion of urban areas is expected to require 100 million ha of additional land by 2030, most of this urban expansion will occur on prime agricultural land because cities were located near their food supplies before modern transportation systems and global food trade, according to Cassman. He writes:

> Humanity is in a race against time to ensure global food security on a planet with limited supplies of arable land, water, and low-cost energy resources, and a rapidly growing human population. Biotechnology and plant molecular sciences provide critical tools for meeting the challenge of food security, but they are not silver bullets. Achieving food security and protecting natural resources will require scientific breakthroughs and technology developments from a large number of basic and applied disciplines.[32]

This connection between energy, water, and food security is worth a bit more discussion as there is increasing concern about our ability to feed an increasing population and provide the energy needed to sustain economic well-being.

According to Margaret Catley-Carlson, vice chair of the World Economic Forum (WEF) Global Agenda Council on Water Security, water is tied to everything:

> Water infuses not only our ground beef patty, lettuce, cheese, pickles, onions, ketchup, and sesame seed bun, but also the bag and packaging in which that hamburger is provided, the building in which it was grilled, the energy to cook it, and the financial system that lent the franchise capital.[33]

More complicated still, according to the WEF, is the fact that these various issues are all highly interlinked, and solutions to one can in fact worsen another: "It takes one liter of water to grow one calorie. This means that a near doubling in food production will not be sustainable without significant – and perhaps radical – changes in agricultural water use."[34]

The German government is evaluating a series of solutions across the food, water, and energy spectrum, and suggests the nexus should be a major focus of the United Nations, as well as the biggest world economies. In a piece titled "A Test Drive for the Rio Goals," the German government says we should ask: "What future do we want for the interrelated sectors of water, energy and food? What are possible goals, targets and actions, indicators, etc.? Which open questions and challenges, such as data availability, persist? What are the main knowledge gaps?"[35]

At a meeting hosted by the German House in New York City, Georg Kell, executive director of the UN Global Compact, said there is a great deal of potential for the corporate sector in implementing nexus approaches. However, he said there needs to be incentive structures which award "front runners" and could help to bring promising initiatives in the corporate sector to scale.

This should be a cue to innovators. There is hunger from both governments and the private sector for accelerators that address the food, water, and energy nexus. But solutions cannot remain in silos. They cannot address solely water, solely energy, or solely food. The future of twenty-first-century water technology lies in comprehending all parts of the nexus and developing innovative technology solutions.

Notes

1 www.rechargecolorado.org.
2 Ibid.
3 World Economic Forum Water Initiative (2011) *Water Security: The Water–Food–Energy–Climate Nexus*, Washington, DC: Island Press.
4 "Nestle's 'Mr. Water' More Worried about H2O than CO2," Reuters, 12 June 2012, www.reuters.com/article/2012/07/12/us-nestle-water-idUSBRE86B0QT20120712.
5 Adapted from http://coloradowaterwise.org/resources/Documents/101311_GEO_Water_Energy_Nexus.pdf.
6 William Sarni (2011) *Corporate Water Strategies*, London: Earthscan.
7 www.google.com/intl/fr/corporate/datacenter/efficient-computing/efficient-data-centers.html.
8 www.google.com/intl/en/corporate/datacenter/efficient-computing/water-management.html.
9 www.google.co.uk/about/datacenters/gallery.
10 www.google.com/about/datacenters/inside/efficiency/cooling.html.
11 *Op. cit.* note 8.
12 www.thegreengrid.com.
13 Ibid.
14 www.rivernetwork.org.
15 Ibid.
16 www.rivernetwork.org/water-energy-nexus.
17 www.energy.ca.gov/2005publications/CEC-700-2005-011/CEC-700-2005-011-SF.PDF, page 7.
18 Sandia National Laboratory, *op. cit.* note 5.
19 www.sbc.slb.com/Our_Ideas/Energy_Perspectives/Winter12_Content/Winter12_Sustainably.aspx.
20 International Energy Agency (2011) *World Energy Outlook 2011: Special Report – The "Golden Age of Gas"*, Paris: IEA.
21 Ibid.
22 Adapted from www.nrdc.org/globalwarming/watersustainability/files/WaterRisk.pdf.
23 Adapted from International Energy Agency, *op. cit.* note 20.
24 CitiGroup (2011) *Citi Water Sector Handbook*, May 24.
25 www.imagineh2o.org/prizes/2010%20Prize.php.
26 Ibid.
27 "Water uses Energy uses Water," www.designecology.com/blog/?p=55.
28 Ibid.
29 http://nabc.cals.cornell.edu/pubs/nabc_20/NABC20_Part_3_4c-Cassman.pdf.
30 Ibid.
31 Ibid., page 172.
32 Ibid.
33 World Economic Forum (2011) *Water Security: The Water–Food–Energy–Climate Nexus*, www.weforum.org/reports/water-security-water-energy-food-climate-nexus.
34 Ibid.
35 "A Test Drive for the Rio Goals," 16 April 2012, www.water-energy-food.org/en/news/view__519/a_test_drive_for_the_rio_goals.html.

Part III

Building the twenty-first-century water industry – ideas, money, and commercialization

Only he who builds the future has the right to judge the past.

(Friedrich Nietzsche)

What should the twenty-first-century water industry look like? What models should we look to for ideas and pathways to success? That may sound like a question best posed to a global think tank, NGO, public policy leader, or a business focused on addressing water-related risks. However, you may be surprised how much influence a local business person can have in helping their business become water-smart, and in turn serving as an example in their city and beyond.

I (Pechet) sat in on a meeting between real estate owners and managers and the new mayor of Tucson, Arizona, in early 2012. I was surprised and excited to hear, without any prompting, the new mayor note that water conservation and management were top priorities of his administration. If you were in that room, you had a voice not only in advancing your business's water practices, but influencing those of Tucson. Moreover, because over 80 percent of America's water systems are publicly owned and managed, those cities see one another as collaborators and are willing to share and borrow ideas collaboratively. If a smart water management practice works in Tucson, a forward thinking and water-stressed city, other cities are likely to consider it. Anyone in that room had the opportunity to start the spread of water innovation like a domino-effect from their business, to their city, and perhaps beyond.

Access to water is a competitive advantage – for business and for countries.

We covered the business value of water in Part I (brand value, business continuity, social license to operate, and as a driver for innovation). It is also a competitive advantage to the public sector for several reasons, one of which is economic growth. No water, no energy, manufacturing, or agriculture, and the private sector pays attention to the availability in a state or community of water to sustain their businesses. In the same Tucson meeting, a publicly traded mining company presented their development plans, which offered significant economic impact on the region, carefully noting their water plans for access to supply and management of takeaway wastewater. According to the USEPA:

- Every job in water and sewer creates 3 new jobs;
- $1 invested in water and sewer leads to $6.35 in long-term private output;
- $1 billion in water infrastructure creates 40,000 jobs;
- $1 billion in water efficiency creates 22,000 jobs; and
- $188 billion to upgrade US water and wastewater infrastructure would generate $265 billion in economic activity and 1.9 million jobs.

Investment in water is smart business for both the public and private sectors.

Access to clean water and sanitation has been a precursor to flourishing economies and business environments. As we consider the ingredients for a twenty-first-century water industry, we can borrow from successes of the past, such as Petra, Rome, and Jerusalem.

The dust-covered stone walls of the ancient city of Petra, Jordan (remember the setting for the *Indiana Jones and the Last Crusade* movie?), don't look like an example of a game-changing center for water innovation. Yet, this smart city of the past can inform our water future. Devising a water delivery system for Petra meant figuring out how best to pump water efficiently and abundantly to a far-off place in the desert – not an easy task. By embracing, creating and understanding advances in engineering and delivery systems, however, the water flowed and the desert city thrived, catalyzing economic development not only within the city but beyond its walls to each place touched by the impact of Petra's bustling trade post.

It all began in 300 BCE when Petra was growing as a commercial hub with exceptional water infrastructure to support economic growth – a virtuous circle of water infrastructure to support growth and attract industries and economic activity funding infrastructure. Water demand grew in step with population growth.

The city is located between the Red Sea, Gaza, and Damascus. At the time, trade routes connected the Orient via the Mediterranean with the Middle East. Because of its location, Petra became the intersection for global trade and, along with it, traders and travelers. Fortunately, traders carried with them more than their wares; they also had insight and knowledge of water technologies from their own cultures. Borrowing the "best practices" from these far-off lands allowed engineers in Petra to design and implement a renowned water delivery system that lasted for centuries.

Petra is a shining example of water innovation and investment driving economy and culture, even in a water scarce environment. Petra's water history emphasizes foresight and collaboration (as we discussed, the potential of "collective action"). Creating an innovative water delivery system in the middle of the desert required foresight into the future benefits: commerce, economic improvement, and higher standards of living for its population. And, collaborating with others to develop a best-in-class system took things beyond mere understanding of the issues into full-scale deployment.

Let's take a look back in greater detail.

Author Charles R. Ortloff explains that piping networks were employed to promote stable water flows and that "sequential particle settling basins" were used to purify supplies.[1] Pressurized systems increased flow rates and leakage concerns were considered and addressed by matching the spring supply rate to the maximum carrying capacity of the pipeline:

> This, and other demonstrations of engineering capability in hydraulic system design indicates a high degree of skill in solving complex hydraulics problems to ensure a stable water supply and may be posited as a key reason behind the many centuries of flourishing city life.[2]

Many credit not just the rise of Petra but the rise of the broader Roman Empire to the sophisticated water aqueduct system it devised. Petra developed innovative approaches to water management which was expanded upon during Roman rule of Petra. The Romans constructed aqueducts to service the cities within their vast empire. This was part of its empire-building plan: increase the quality of life through superior infrastructure and services to attract allegiance and keep rule.

The city of Rome itself, of course, had the largest number of aqueducts enabling it to serve fresh drinking water and supply its famous fountains and baths in spectacular strength and cultural display. Moreover, the Romans developed a sewer system and redirected the used "gray water" to remove waste. Clearly, a much higher quality of life was achieved there than in the outposts of civilization where even an accessible water supply was a luxury.

Rome's water system was built over a period of several hundred years and eventually covered more than 500 miles of piping networks. Taken into account were the materials to be used – clay rather than lead was utilized because people quickly learned of lead's toxicity; the aqueducts' positioning – most aqueducts were built below ground to keep water free from disease and enemy attack; and gravity – arches were built to maintain the pitch of the aqueduct and to keep the water flowing over long stretches of irregular terrain.

In short, the Romans built one of the great wonders of the world, and the people of Petra (among others) borrowed generously from those technological advances. Much of the design thinking is still employed today when water systems are planned and implemented in modern townships and cities across the globe.

There are several factors that contributed to the success of the Roman aqueducts; careful planning, *long term investment* (we will discuss the need for long term investment in water infrastructure later) and routine maintenance It is worth noting the Aqueduct of Segovia as an example. This is one of the best-preserved Roman aqueducts in Spain. So well-built was the aqueduct and so studious its maintenance through the Middle Ages that it functioned as a viable water delivery system well into the twentieth century.

Beyond advance planning of system and designs, Roman engineers accounted for externalities and variables that still challenge modern water planners. They knew the value of water.

Building the aqueducts, however, was only part of the equation. Rome's engineers also detailed a comprehensive system of regular maintenance to repair accidental breaches, to clear the lines of debris, and to remove buildup of chemicals such as calcium carbonate that naturally occur in the water. The methods of all this construction and planning were written down and described by Vitruvius in the *De Architectura* circa first century BCE. Rome's engineers also understood the impact of illegal piping, water theft, and aqueduct breaches, and managed these issues.

Historians note that Roman aqueducts were extremely sophisticated constructions, built to remarkably fine tolerances. One aqueduct alone, for example, could transport up to 20,000 cubic meters – nearly 6 million gallons – a day, powered solely by gravity. The combined aqueducts of the city of Rome supplied around 1 million cubic meters (300 million gallons) a day. This was an incredible feat for not only its time, but for any time. New York City, by comparison provides 1.2 billion gallons of drinking water per day to its 8 million residents (using a similar system of gravity, it should be noted, along with a protected watershed in upstate New York – engineering coupled with smart, forward-looking public policy).

Another example of "ancient" technology comes from ancient Jerusalem and gray water plumbing.[3] Ancient Jerusalem applied gray water plumbing techniques that are again today generating renewed interest. Located at a high elevation and away from plentiful sources of surface water, the ancient city of Jerusalem has relied on aquifers for nearly 15,000 years, and still has a network of wells and tunnels that date back to the twelfth century BCE. As the city grew and evolved, so did its water use and reuse systems. Sink water was conserved in basins and used to flush waste, much like modern sewers, but also saved to water gardens while particulates were filtered to provide fertilizer for surrounding fields.

Today, water reuse, from similarly basic methods to advanced treatment systems, is one of the fastest growing sub-segments of the water industry.

The use of ancient technologies such as gray water plumbing is returning as we recognize how foolish it is to waste high quality water for low quality needs such as flushing toilets or landscape watering. We are also creating innovative water solutions through a "mash up" of seemingly unrelated technologies. It should be no surprise that modern innovation in the water industry will leverage a wide range of technologies – some of these coming from unexpected places. For example, the mobile phone, and its associated technologies, may provide a key to addressing water and sanitation needs in developing countries.

These ancient cities and cities of today share significant water challenges. Although not an easy task, technology innovation can overcome these challenges.

As William Blake wrote, "Great things are done when Men and Mountains meet; This is not done by jostling in the street."

Notes

1   Charles R. Ortloff (2005) "The Water Supply of Petra," *Cambridge Archaeological Journal.*
2   Ibid.
3   http://webecoist.momtastic.com/2009/01/25/ancient-green-architecture-alternative-energy-design.

Chapter 7

The ideas

Who generates the innovative ideas to solve some of the most challenging water issues? Who will craft the solutions to build the twenty-first-century water industry?

Historically, water innovation has come from practitioners within the public and private sectors of the water industry. They are composed of service providers, a range of technology companies, and those engaged in public policy issues. Despite significant challenges, these stakeholders have cultivated reliable water systems across the globe, for developed and developing economies alike. These challenges include constraints imposed by limited government budgets (the relentless pressure to reduce expenditures) and within the private sector confronted by very poor investment payback timeframes. Neither situation is conducive to risk taking and innovation.

The limitations on risk and investment in innovation contrast sharply with the current size and projected growth of the water industry (which now exceeds $500 billion in annual sales, according to Lux Research[1]). Based upon the scale and potential of the industry, one might assume a burgeoning portion of the industry composed of new entrants pushing new ideas and new solutions to compete with incumbents in the new water economy. Instead, adoption of innovation feels surprisingly slow, and we remain early in the innovation adoption curve of the industry. The figures on capital investment in early-stage water opportunities support that sentiment. Investment in innovation across sectors has ranged around $25 billion annually in recent years; however, less than $250 million, or about one percent of all angel and venture capital investment, funds water innovation.

Why? If this is the case, then how will we come up with new ideas? Or better yet, how can we change it to build the capacity for new solutions to water needs? The solution lies, in part, in opening up innovation to a broader group of stakeholders.

One example of a pathway for innovation in the water industry is Imagine H2O.[2] Imagine H2O was started to spark innovation and help innovative solutions reach customers. The founders saw the organization as one piece of a broad effort to reverse a vicious cycle in water innovation. A pervasive lack of awareness of opportunities in water, and how to capitalize on them, leads to lower than needed supply of new innovation. Lack of motivated, educated capital to carry innovation forward, from both government and private sources, provides limited allure to innovators, and fails to grease the wheels for those attempting to capitalize entrepreneurial ventures. Common obstacles impeding customer adoption of innovation, such as limited access to big customers, like major public utilities, and the inherent risk-aversion to new methods of managing water, further stifle the innovation cycle. The idea was to address each piece of this vicious cycle in order to help convert it to a virtuous cycle.

Focusing on what it calls "The Water Opportunity," Imagine H2O offers competition, cash prizes, business incubation and a water-resources network.

The founders of Imagine H2O are seeking answers to global water questions that no one could provide:

- What does the water market look like in dollars?
- What problems need to be solved?
- Which of those offer the best commercial opportunities?
- With whom should we connect for help, advice and partnership?

In order to find answers to these questions, Imagine H2O decided they would need to mobilize people around the world who were willing to investigate water problems and find viable, commercially sustainable solutions to these problems. If there were competitions that awarded large cash prizes, those people would come forward; and if the contestants were offered more than simple cash prizes and trophies – if they were given real sustenance, such as business incubation – then local economies would be strengthened while the global water crisis is being solved.

They opted for the creation of a nonprofit, armed with the knowledge that nonprofits are in a unique position to unite stakeholders from across all sectors: government, business and the people. Appropriately, Imagine H2O also seeks to create an "ecosystem" of water innovators and resources from around the world to support the new business ventures.

Although the water industry has many non-profits, most charitable dollars flow to water issues in developing economies such as drilling new wells, addressing poor water quality or watershed protection. Imagine H2O offered a market based approach to catalyzing the water economy. Such market-based, capacity building initiatives will be crucial to the development of the twenty-first-century water industry.

Citing historic competitions such as the Orteig Prize, which prompted the success and innovation of Charles Lindberg, Imagine H2O believes nothing can motivate and generate creative thinking quite like good old-fashioned competition. A proven tactic in sectors across business, offering a competition appeals to those with an entrepreneurial spirit. It can be enough to nudge a burgeoning idea into reality, and, when it comes to solving the global water crisis, it can be a lifeline into communities where water issues are great and financial support is not. Imagine H2O wants to 'harness the power of competition, capitalism and entrepreneurship, instead of relying on philanthropy and policy-based solutions' in the undeveloped and developed world.[3]

Open to anyone and everyone around the world, the Imagine H2O annual competitions change focus each year and aim to support as many viable business ideas as they can. The non-profit also sees the value of competition outside the business arena. In the future, there will be competitions in the areas of science, youth innovation, and policy. These are all areas where Imagine H2O wants to help "turn great ideas into real-world solutions that ensure available clean water and sanitation" around the world.

Led by successful innovation prize operators like the X-Prize Foundation, competitions in business planning, science and technology have grown in popularity. Imagine H2O uses its prize as a means to an end – finding worthy ideas to accelerate to real-world solutions that reach customers. It now partners with several existing prizes to feed its water innovation accelerator program. Guiding Imagine H2O, other market-based approaches to water issues, and the entrepreneurs and innovators who drive them, is the belief that the large struggle for water is a large business opportunity. By keeping the focus of each competition solely on water and ensuring that well-developed countries are not excluded as potential users of the new products and services being presented, Imagine H2O has been able to attract a wide range of contestants, and bring winning companies access to capital markets and customers.

The winners are not only given cash prizes but are also supported with in-kind services in the areas of accounting and law, and given resources and contacts in other areas that can be hard for a new business to obtain. But Imagine H2O doesn't believe supporting only the winners will create real change around the world quickly enough, as time to put effective measures in place is running short. The nonprofit provides seminars, workshops, business feedback, networking and team-building opportunities, and guidance to every single contestant.

So far the competitions have helped to make water savings from businesses in the areas of irrigation, rainwater storage, water-based software programs and the energy–water nexus.

Perhaps the most vital aspect of Imagine H2O is the organization's drive to create and provide "an ecosystem of stakeholders to the next great water innovations."[4] An ecosystem implies dynamic, interconnected relationships, not static operations. With this as the model, Imagine H2O uses its nonprofit status to position itself in the middle of this intricate web of business resources, helping guide both business and resources as they look for appropriate connections. Water is a global issue with very local implications. As the thread that weaves Imagine H2O's web of interconnectivity together, it is, in more ways than one, an underpinning to its success.

Increasingly, cities, countries, and universities are launching initiatives to spark water innovation and to capitalize on the potential growth of the water economy, and its multiplier effect on the broader economy. In the past few years, Milwaukee, Wisconsin launched The Water Council,[5] and Colorado State University in Fort Collins, Colorado launched The Global Water Initiative,[6] to name but two.

There is a rising tide to drive innovation in the water sector by identifying and nurturing start-ups.

The start-up

How do you go from a need, via an idea, to innovating a solution that has an impact?

Picture this:

> Five boys, each carrying a five-gallon water jug, spill from the back of a bright red pickup and crowd into a shed no larger than a walk-in closet. The boys hand their jugs over to a young man and woman waiting inside. "It's a family event," comments volunteer Shane Nichols as he collects the jugs. The sky-blue shed, called the water hole, is connected to a 5,000-gallon mobile tank that stores mountain spring water hauled in from sixty miles away. Angelica Martinez – a high school senior who volunteers here with Nichols every Saturday – places the boys' plastic jugs under a row of spigots and turns on the water. The shed's thin metal walls reverberate with the sound of water falling into plastic . . .

This isn't a scene out of some developing country. This, as reported by John Gibler in *Terrain* magazine, is taking place in California.[7] "The water hole is Alpaugh's sole source of safe drinking water. Three times a week volunteers unlock the shed door and help residents fill jugs to take back to their homes," Gibler reports.

Arsenic concentrations in groundwater impact the town of Alpaugh in the Central Valley, California, and many other areas of the United States – so much so that the USEPA has lowered the acceptable concentration of arsenic in water to 10 parts per billion (ppb) from the longstanding 50 parts per billion.[8]

In the Central Valley, the arsenic concentrations in some wells exceed 86 parts per billion. Few places in the world have this high a concentration of arsenic and high concentrations of arsenic can result in a variety of health problems.[9]

It should be noted that arsenic is commonly found in water. It's when concentrations exceed the regulated concentration that health concerns increase. Arsenic in groundwater can be both naturally occurring and from man-made releases. Arsenic can occur in rock formations and is released into groundwater as it dissolves over time.

In general, compared with the rest of the US, western states have higher concentrations of arsenic than the USEPA standard of 10 parts per billion (ppb), the agency reports. Parts of the Midwest and New England have some systems whose current arsenic levels are greater than 10 ppb, but most concentrations range between 2 to 10 ppb, according to the USEPA.

The water hole in Alpaugh is necessary for two reasons: it replaces the two main water wells, which broke some years ago, and it supplies much cleaner water than the wells were able to deliver. Water samples taken at the time of the wells' breaking showed the presence of both coliform and arsenic in the community's water source. Ironically, Alpaugh sits next to the largest and most expensive water system in the in the world. It's used to supply the agricultural region of the San Joaquin Valley of California. Over time this changed, and the area became reliant on groundwater supplies. "Early irrigation projects used pumps to pull water from the underlying aquifers. The pace of agricultural expansion in the first decades of the 20th century over-drafted these aquifers, causing the valley floor to sink as much as thirty feet in some areas," according to *Terrain* magazine.[10] "Over-drafting" or over pumping of aquifers can cause arsenic levels to spike in water wells.[11]

In other places around the world arsenic contamination is also prevalent. The most prominent of places subject to arsenic contamination is Bangladesh.

According to a website devoted to raising awareness about arsenic contamination in the area,[12] the problem in Bangladesh occurred when there was a move to expand and improve the water delivery system: "In the face of the heavily contaminated surface water, the Indian and Bangladeshi governments and foreign aid agencies took action in the 1970s and '80s in order to provide the population with potable drinking water."

These governments and agencies dug over a million public pipewells to shallow aquifers, providing the population with water that was believed to be clean and safe. A pipewell or tube-well consists of a tube of 5cm diameter inserted into the ground and capped with an iron or steel hand pump. The majority of the wells were installed by the United Nations Children's Fund, UNICEF. Ironically these wells led to the largest arsenic contamination crisis in the world. In later years, privately sunk wells became more prominent, eventually outnumbering public wells. Currently more than 90 percent of the drinking water in Bangladesh derives from aquifers less than 300m deep, with most aquifers less than 100 meters deep.[13]

In the 1980s health problems associated with high arsenic concentrations were discovered in West Bengal. Despite these clues, the arsenic problem remained unknown until a 1993 examination of well water in the Nawabganj district led to the discovery of high arsenic concentrations. The number of people drinking water from wells contaminated by arsenic is staggering. Twenty-seven percent of the shallow wells have concentrations of arsenic exceeding the Bangladeshi standards of 50 micrograms per liter (equivalent to ppb), which is five times the World Health Organization's standard of 10 micrograms per liter. It is estimated that up to 30 to 35 million people in Bangladesh and 6 million people in West Bengal are exposed to concentrations of 50 micrograms per liter of arsenic in their water or more.[14]

The crisis in Bangladesh and West Bengal highlights the health problems associated with arsenic poisoning. Skin problems such as keratosis, pigmentation and de-pigmentation, and skin cancer are widespread. Patients who have been exposed for a longer period of time have developed internal organ problems such as cancer, and many more have died. The arsenic poisoning was first identified in West Bengal in July of 1983. Since then, the numbers of people exposed to high arsenic concentrations has increased.

This is where water technology and innovation can create massive social effects and provide financial opportunities. Indeed, with such catastrophe comes opportunity on a global scale.

No other water issue today may present as big an opportunity to improve water quality than treating arsenic contamination.

John Schroeder figured this out, and he went on to build a water company founded on technology developed at Rutgers University. Schroeder was approached to run one of the companies spun out of Rutgers' Ecocomplex, an incubator business focused on technology start-ups. In the case of Hydroglobe, which was originally named Net-O-Filter, Schroeder saw something big: a way to purify water from arsenic and lead, among other contaminants.[15]

"As I got into it, I found out that other people were already using the same process we were," Schroeder recalls. He explained that the company that would eventually become Hydroglobe was using iron-based material in liquid or solid forms as a purifier.

He went back to the lab. In development was a titanium dioxide-based absorbent. It was a new way to treat water with arsenic concentrations – in powder form.

He immediately got the patent process started and attracted investors. Then the orders came.

"We had some small customers and orders for filters that used iron pellets," he says. Then along came Brita, the water filter company. "Brita said 'this is a really neat product,' and they decided to incorporate it into one of their end tap filters."

That got Hydroglobe jumpstarted. But it was just the beginning.

He began giving presentations, talking to the USEPA, and generally getting known in the industry. "I'd speak with anyone who could help validate the product," Schroeder says.

Several years into the company's development, with some venture capital and a small sales staff, Schroeder began reaching out to larger companies that might be interested and found Dow Chemical. "Dow thought this [Hydroglobe's technology] was the best things since sliced bread and I got them to license the product," he says.

The story gets even better.

Another sales call he made was to Graver Technologies. A 30-year-old company, Graver develops, manufactures and markets a wide array of products and services for the power generation, food and beverage, drinking water, pharmaceutical and chemical markets. "They listened to my sales pitch and the next day I got a call from the head of research and development (R&D) asking if I was interested in selling the company," Schroeder says. He was, and he did.

It turned out that they didn't just want to buy the product; Graver wanted to keep investing in the company. "Every year it's another record," Schroeder, now the president of Graver Technologies, says. "We make a nice environmental product that is a solution to a lot of problems."

There were many challenges along the way to Schroeder's success. For one: the idea that Hydroglobe solved a problem and helped companies abide by new regulatory requirements was actually viewed as a negative by investors. "People will just go around [the regulations]," was the mentality investors had.

"When the USEPA sets a new standard it is usually based on a one in a million chance of anything bad happening. In this case [with arsenic] the chances of something bad happening are one in 300," Schroeder says.

Arsenic contamination is a big problem. It's also an example of a problem that the customer must address for regulatory and health reasons. However, many technology solutions exist to water problems unrelated to regulatory issues (such as water efficiency) and typically customers do not prioritize purchasing those solutions. Most successful water start-ups target "must-spend" customer categories.

Interestingly, Mr Schroeder's start-up story follows an uncommon path. Mr. Schroeder approached the industry from the outside and succeeded. Most water entrepreneurs arise from within water agencies and utilities, or from large water companies.

Mr Schroeder joined Imagine H2O as an early advisory board member because he recognized the need to develop a pathway to replicate his story for other innovators. Again, this is what is needed to increase innovation in the water industry: an influx of new people, ideas, and methods.

What makes a water tech start-up successful? Attributes include:

- the ability to craft and communicate a clear and compelling value proposition to customers *despite* the current economics of water;
- access to consumers and decision makers;
- creating customer value beyond the price of water, from solutions which use less energy, use fewer or no chemicals for treatment, generate less waste materials or recover resources from wastewater, or generate higher crop yield; and
- a proven management team.

A small sampling of successful start-ups (not meant to be exhaustive) illustrates these points:

- *Emefcy*[16] – The company has developed a wastewater treatment solution with the ability to generate electricity through a microbial fuel cell (MFC).

The technology can treat industrial wastewater from dairy products, pharmaceuticals and food additives. The MFC solution produces less sludge than conventional processes. The core product is the "megawatter," a wastewater treatment product using an MFC. MFCs are biologically active fuel cells that produce electricity from the degradation of wastewater treatment with near zero power consumption and a reduced sludge yield of about 80 percent.

- *HydroPoint Data Systems*[17] – Their WeatherTRAK system is comprised of a controller and scheduling software deployed at the site. This system periodically accesses about 14000 weather stations in the US and the modeling software then validates local weather down to one square km before scheduling updates. It is the developer of a satellite-based smart irrigation system that continuously analyzes local weather to optimize and control water for landscape irrigation. This reduces pollution runoff, lowers water bills and reduces the potential of property damage from overwatering.

- *Puralytics*[18] – The company has developed a patent pending water purification technology that uses a light activated photo catalyst nano coating. Water is purified by five simultaneous photochemical reactions breaking down organic compounds, reducing and removing heavy metals and sterilizing microorganisms. There are no chemical additives and 100 percent of the water is purified.

- *Voltea*[19] – Voltea's desalination technology, CapDI (capacitive deionization), desalinates brackish water at a lower economic and environmental cost than any other available technology. CapDI is a simple and innovative way to remove dissolved salts from water. The company is focused on developing flexible and economical water treatment technologies from water softening systems for laundry machines to cooling towers, desalination of brackish groundwater and eventually commercial greenhouses. The technology is a scalable desalination technology using membrane capacitive deionization.

- *APTwater*[20] – The company "has developed a revolutionary technology that can remove nitrates and other oxidized contaminants from agriculturally impacted source water without producing waste." Instead of selling its technology, the company developed a business model to re-commission abandoned wells, and to sell water from those wells treated with its technology.

- *TaKaDu*[21] – The company provides software for water utilities to monitor their networks, detect leaks and address inefficiencies in delivery. Rather than selling based only on the value of avoiding water loss, the company provides a value proposition to its utility customers to better maintain its costly infrastructure of pipes and pumps underground.

- *Water Health*[22] – The company is a decentralized water utility that delivers access to clean drinking water in underserved communities in India, Bangladesh, Ghana and the Philippines for less than US$10 per person. Rather than focusing on breakthrough technology, Water Health focused on the partnerships and business model to gain distribution to customers in remote locations, and to minimize maintenance requirements on those systems.

- *Ostara*[23] – The company recovers nutrients (nitrogen and phosphorus) from municipal and industrial wastewaters and transforms them into fertilizer.

It is also worth revisiting a company highlighted in *Corporate Water Strategies*[24] to see how far they have come in the past two years, and as an example of what success looks like. The Water Initiative (TWI)[25] describes itself as:

a team of leading global business executives and renowned scientists who develop and deploy POD or POU water systems to fit local conditions. TWI engages local communities to customize and create comprehensive and sustainable technology solutions, which effectively remove water contaminants such as pathogens (bacteria and viruses), unsafe levels of inorganic materials (such as arsenic and fluorides) and other harmful chemicals and contaminants of concern.

TWI began installations of POD devices to remove excessive arsenic, pathogens, and other contaminants in accordance with an initial state contract to install over 31,000 devices throughout the State of Durango, Mexico.[26] According to TWI the launch of its POD filtration solution, WaterCura®, is a key first step in the process of showing that centralized water treatment is not the only answer in providing contaminant free and affordable clean drinking water at the point of where it is consumed or used.

Again, according to TWI, this launch follows a November 2011 decision by the State Water Commission of Durango, which awarded TWI initial contracts to manufacture and install over 30,000 innovative POD devices for rural and urban homes to create total clean water solutions for the contamination issues in the area.

TWI's solution was approximately 30 percent of the cost of centralized treatment and 16 percent of other POD proposals submitted in response to the public bidding process. In total, over 50 rural communities and 50 urban communities across four of Durango's municipalities (equating to a population of nearly 150,000) will benefit from safe, clean drinking water through the WaterCura installations.[27]

TWI first entered Mexico in 2007 and has worked over the past several years to customize and create a comprehensive sustainable solution for clean drinking water. Their methodology is to diagnose the different water quality conditions and cultural needs; develop appropriate solutions that are affordable, convenient, effective and trustworthy; and then deploy the appropriate solution.

"We went to Mexico, our neighbor, first because it's a microcosm of the world's worst water contamination conditions," said Kevin McGovern, TWI Chairman.

The government of Mexico has certified TWI's products as the appropriate solution for their water challenges. In fact, earlier work in Mexico by TWI resulted in being awarded "The Global Game Changer" recognition in 2011 by the EastWest Institute (EWI) for its work in developing public–private partnership solutions to address the drinking water contamination issues in Mexico.

"With our historic development and proof of concept, we are taking our unique approach to solve water contamination problems in other countries with clean water programs," said Mike McGettigan, Chief Commercial Officer at TWI.

Intellectual property – the patents

Several years ago, I (Pechet) attended a meeting of entrepreneurs and big water companies, hosted by a law firm. The question from entrepreneurs to big companies was (to paraphrase) "will you steal my innovation?" One would have expected entrepreneurs to press their big company counterparts first and foremost on "how can you get me to market?" Herein lies one of the challenges regarding patents in the water industry: many water entrepreneurs miss commercialization opportunities because they are focused on intellectual property protection to the detriment of adoption of their solutions.

Although that phenomenon might apply across many industries, several nuances of the water market exacerbate the issue. In many industries, customers aggressively pursue, and rapidly adopt, breakthrough technologies. Although there are many exceptions, broadly speaking, customers of water technologies want a full solution that delivers the water or removes the wastewater, provided in a manner that they can trust. Most water customers do not seek out piecemeal technologies that are part of the solution they need. They seek service providers who deliver the solution, and their willingness to pay for better technology is dampened by the difficulty of isolating the effect of the technology on the ultimate, integrated solution. For example, one innovative company created a biological dosing solution to treat sewage in municipal pipes before it reached a treatment plant. The company argued that treatment plants would reap numerous benefits, including reduced energy requirements for treating and lower sludge output. As with many environmental solutions, the company faced the challenge of isolating the effect of its solutions from the myriad variables in a dynamic environmental system. However, in a theme common throughout water innovation adoption, customers first wanted to know "who else like us uses this technology?" Most water customer categories are highly fragmented, and customer references are a necessity.

These factors lead to the suggestion that, all else being equal, intellectual property may be valued less in the water market than in other markets. That suggestion has implications for how we reach our vision of the twenty-first-century water market. It implies that many worthy technologies lie latent, rather than reaching customers that need them. It does not suggest there is limited value in innovation in the water sector, only that one key to building a better water future may be unlocking good ideas from over-protection. Interestingly, data shows rapidly rising patent filings in water.

The increase in water technology patents perhaps reflects how innovators value the protection of their ideas. According to a 2010 report by Foley &

Lardner, water-specific patents increased 52 percent from 2008 to 2009, and are continuing to grow.[28] Whether or not these ideas get funded is a separate matter. And whether these funded ideas are translated into commercialized products is yet another matter.

Building the twenty-first-century water industry will take all three components to innovation – ideas, funding, and commercialization.

As Lang McHardy, an intellectual property expert with Vested IP says:

> The basics of how to move and process water have been around for centuries. People tend to think that patents are only suitable for "flash of genius," "paradigm shifting" technologies. The reality is that most patents are actually directed to relatively small incremental improvements to old technologies. Building a cohesive portfolio of incremental-improvement patents around a common approach to improving a particular technology is a great way to create value. Incremental improvements are also easier to sell into the water market (where utilities tend to shy away from big dramatic changes). So, it's critical for entrepreneurs and engineers to avoid the trap of assuming that a particular innovation is too trivial or insignificant for a patent.[29]

Patents have an important role to play in moving ideas to the commercial stage and it is worth providing an overview of the types of patents (Box 7.1), trends in patents and some perspectives.

While patent definitions are relatively straightforward, approaches to patents are not and, as you would expect, opinions vary. Additional insight on the value of patents and trends from Lang McHardy from Vested IP is summarized below:

- One of the first things to understand is that "patent strategy" means something very different for a start-up company as compared with a large, established company. Many start-up entrepreneurs make the mistake of looking at patents from the perspective of a large company. To a large company, patents are predominantly bargaining chips to be used in cross-licensing negotiations with rivals, and to protect market share through litigation. As a result, many start-up entrepreneurs mistakenly assume that, because they don't have the resources to build enormous war chests of patents or to enforce even a single patent, it's not worth spending the money to get them in the first place.
- To the extent that the goal of a scalable start-up is to build a company that can be sold to a larger company or to the public market, patents are an essential part of the package. The ideas, innovation and know-how that go into technology is all intangible. Patents are a way of making those ephemeral, intangible concepts into assets that can be sold, licensed, bargained with, etc. In an exit, having patents will typically make the company more valuable than not having any. Even in the

Box 7.1 A basic overview of patents[30]

A patent is a legal document that defines a set of exclusive rights to a new technology, product or service. The exclusive rights are granted to the patent owner for a limited amount of time and can be leveraged in a variety of ways to support a business strategy and add value to a business.

Patents have many strategic uses. Patents can be used to create a legal barrier to competition, to establish a portfolio of assets that can be used to generate revenues through licensing or IP transfers, or to augment the value of a business for purposes of raising seed or venture funding.

What types of innovations are patentable?

Utility patents can protect inventions that are novel (new), non-obvious, and useful, including the following types of subject matter:

- *Process or method* (e.g. method of making or using a product or computer-based processes).
- *Machine* (something with moving parts or circuitry).
- *Article of manufacture* (such as a tool or another object that accomplishes a result with no moving parts, such as a new hammer).
- *Composition of matter* (such as a new pharmaceutical, food, or toothpaste).
- An improvement of any of the above items. Most patents are for incremental improvements made to known or pre-existing technology – thus typically evolution rather than revolution.

Design patents may be granted for any new, original and ornamental design for an article of manufacture and protects only the appearance of the article and not its structural or utilitarian features. In general terms, a *utility patent* protects the way an article is used and works, while a *design patent* protects the way an article looks. Both design and utility patents may be obtained on an article if invention resides both in its utility and ornamental appearance.

Even if the invention falls into one of the above categories, there are certain subject matters that cannot be patented, including mathematical formulas, naturally occurring substances, abstract ideas, or laws of nature.

What does a patent contain? It must include at least one claim. Also, typically:

- background of the invention;
- summary of the invention;
- detailed description of the invention; and
- (optionally) examples (actual or prophetic).

The patent claims define the exclusive property right provided by a granted patent. Patents include one or more independent claims and may include one or more dependent claims that further narrow the independent claims. Dependent claims are helpful in the event the broader independent claims are later found to be invalid.

unfortunate case that the company fails, the patents may be the only remaining assets that hold significant value.

- Whether patents are really valued in the water industry or if customer adoption is valued more heavily – in my opinion, both are necessary for a technology-focused water company. It's certainly not an either/or situation. It's very much possible to pursue both simultaneously. Of course, this assumes that we're talking about a company whose main innovations are in technology, as opposed to a company whose innovation/uniqueness is in project management, sourcing or financing. If the innovation and development of new technology is not part of the business model, then patents don't make sense.

- Water start-ups are very similar to other clean tech start-ups; they get VC funding from many of the same firms, and they're looking to exit by getting bought by many of the same big companies or IPOing to the same public market. VCs recognize that patents will be important in an exit, and so they'll often expect to see some as part of the overall strategy (even if many of them aren't any good at evaluating how good a company's patents actually are).

- Also similar to other clean tech fields (e.g. "alternative" energy sources), modern companies are mostly refining technologies that have been around a long time. The basics of how to move and process water have been around for centuries. People tend to think that patents are only suitable for "flash of genius", "paradigm shifting" technologies. The reality is that most patents are actually directed to relatively small incremental improvements to old technologies. Building a cohesive portfolio of incremental-improvement patents around a common approach to improving a particular technology is a great way to create value. Incremental improvements are also easier to sell into the water market (where utilities tend to shy away from big dramatic changes). So, it's critical for entrepreneurs and engineers to avoid the trap of assuming that a particular innovation is too trivial or insignificant for a patent. It's obviously important to strike a balance between managing patent costs and pursuing patents for "everything," so in addition to the question of "is it patentable," one must also evaluate whether the patent will likely be worth anything to anyone else (both with and without the company's other technologies).

- Regarding trends, patents for pure software (i.e. software that only manipulates information rather than controlling some physical apparatus) are increasingly popularly sought and at the same time are increasingly difficult to get. This reflects the fact that software is eating the world. Patent applications (which are published generally 18 months after filing) are generally lagging indicators of broad technology trends – by the time you spot a trend in patents, it's probably already a trend in technology development. Many water companies have novel software (e.g. supply/demand/usage analytics), and should pursue patents for it when it is truly novel, but the evaluation of the value of a potential software patent may demand a slightly higher bar given the increased difficulty in getting them allowed.

- As for the recent changes to patent law, they don't really change much when it comes to "best practices" for building a patent portfolio. The idea that the "first-to-invent" could actually win the patent was largely a red herring leading to a false sense of security. In reality, disputes between two inventors claiming to have invented the same thing are almost always won by the first-to-file. Hence, the best-practice was, and remains, to file patent applications as soon as possible after identifying a valuable and patentable invention.

- You might see some question about protecting technology development as Trade Secrets vs. patents – this may be less risky in view of some of the recent patent law changes, but Trade Secrets are relatively difficult to value and sell as assets. This is because the "secret" is rarely as simple as a recipe, they're often not written down or memorialized in any real way, and they are usually dependent on special "know-how" in one or more employee's head. I think patents are a better store of value in most cases.[31]

According to Augie Rakow:

Patenting activity is not uniform across all fields of water technology. The fields of purification, reclamation / treatment and conservation produce 5 to 6 times more patents than the fields of irrigation, desalinization / distillation and metering. In particular, the most active fields are water flow-reducing fixtures and flow limitation valves, water-conserving appliances, industrial waste water treatment, sub-surface hydro generation, ionization- or electrode-purification devices, pool/pond filtration devices, controls for water-treatment systems, water treatment materials, steam distillation, and ground / storm runoff filtration.[32]

The most essential points about patents are:

- many entrepreneurs tend to protect their ideas and patents at the expense of securing funding and commercialization; and
- each situation is unique.

In water, unlike other fields, getting to market is challenging. The sales distribution channels are hard to understand and to penetrate, and customers require long testing periods with lots of proof points and pilot tests. That reduces the value of the underlying technology, and increases the value of product and service distribution.

To quote Doug Henston, entrepreneur and former CEO of Solix, in illustration of these points:

- Investors don't invest in technology itself, they invest in businesses. Engineers have an emotional attachment to the technology, protecting them like they are babies. Moving from technology to business means you need to let go of that attachment. It now belongs to the investors and the company.
- Engineers don't have the perspective they should have. So they may overprotect something simply because they have an emotional stamp, akin to a personal identity stamp, which is not appropriate. Engineers might feel like their tech can save the world, but investors don't want to hear that – they want to hear how something works and how it will make money.
- Operating knowledge should not be patented – it's the "secret sauce" and can't be copied. Patenting makes something public knowledge. Try and protect yourself with non-disclosure agreements (NDAs), hard coding some of the trade secret into something like software that you're using.
- Look at your portfolio: what should be copyrighted, patented, trademarked and what should not. The idea of filing the "perfect patent": you want the product to be perfect and then file patent, and in the meantime someone else beats you to it. Then you've lost out totally. The market moves too quickly now.[33]

Ideas and their patents are the beginning of the journey to commercialization. The next step is finding the money.

Notes

1 www.e-zcoursebooks.com/WaterEnergyNexus.pdf.
2 www.imagineh2o.org.
3 www.imagineH2O.org/about/ourmodel.php.
4 www.imagineH2O.org/about.
5 www.thewatercouncil.com.
6 www.globalwater.colostate.edu.
7 John Gibler (2005) "Not a Drop to Drink," *Terrain*, winter.
8 http://water.epa.gov/drink/info/arsenic/upload/2005_11_10_arsenic_ars_final_app_b.pdf.
9 http://water.epa.gov/lawsregs/rulesregs/sdwa/arsenic/Basic-Information.cfm.
10 Gibler, *op. cit.* note 7.
11 http://southeastfarmpress.com/shrinking-groundwater-supplies-point-much-needed-reforms.
12 www.sos-arsenic.net.
13 www.sos-arsenic.net/english/contamin/index.html.
14 Ibid.
15 Interview with John Schroeder for this book.

16   www.emefcy.com.
17   www.hydropoint.com.
18   www.puralytics.com.
19   www.voltea.com.
20   www.aptwater.com.
21   www.takadu.com.
22   www.waterhealth.com.
23   www.ostara.com.
24   William Sarni (2011) *Corporate Water Strategies*, London: Earthscan.
25   www.thewaterinitiative.com.
26   www.prweb.com/releases/water-contamination/clean-water/prweb9440241.htm.
27   *Op. cit.* note 25.
28   Foley & Lardner LLP (2010) *Water Technology US Patent Landscape Annual Report*.
29   www.vestedip.com and interview with Hillary Mizia for this book, 24 September 2011.
30   Adapted from Rimon Law Group (undated) *Start-up Package*, pages 15–16, http://bit.ly/emXsPT.
31   Interview with Hillary Mizia for this book, 24 September 2011.
32   Ibid.
33   Ibid.

Chapter 8

The money

We have argued that the way we manage water today will not be the way we will need to manage water tomorrow. We have discussed the criticality of improved water technology, policy, pricing and practice. To catalyze the shift in approach in managing water takes money. Historically, where has the money come from? Where will the next generation of solutions find funding?

Water has long been viewed as a unique asset class, dominated by government funding and government buying. To get from where we are today to where we need to be tomorrow, can we rely on rational private capital markets for water solutions? If there is a market failure in water finance, what can we do to correct it to catalyze the twenty-first-century water industry?

In April 2012, I (Pechet) watched a classroom full of future leaders in finance debate the functioning and merits of capital markets in water. Professor André Perold led a class called Real Asset Finance at Harvard Business School, where he had run the finance department for many years.[1] On this particular April afternoon, he taught cases he had written on each organization as a way of exposing some of the school's brilliant financial minds to water as a business. The class had studied other real asset classes, such as oil and gas, timber, and metals, and the discussion provided a framework for analyzing the opportunity for water.

The case studies began with a recent quote from Citigroup's chief economist, Willem Buiter: "Water as an asset class will, in my view, become eventually the single most important physical-commodity based asset class, dwarfing oil, copper, agricultural commodities and precious metals." Water has long been heralded as the oil or gold of the future. Yet the students, whose investing backgrounds ranged across leading banks, private equity firms, and consultancies, offered a mostly negative perspective on water investing. They questioned cheap prices for water and sewer, the inconsistent and somewhat unpredictable role of federal, state, and local governments in any water investment, lack of customer drive, and other common obstacles to making money as a private investor in water. A class of the smartest minds in water loved the idea of investing in water, but struggled with the practical realities of finding ways to earn a return.

As the students debated whether to invest in water (and, if so, how), Professor Perold highlighted an interesting angle to the conversation, and one that underlies a critical component of capital formation in water. This was the first time many students had thought about water at all, no less water as an investment. Professor Perold served on the board of directors of the Vanguard Group, investing over $1.5 trillion in mutual fund assets, and offered an experienced, high-level perspective. "Isn't this exactly what you want?" he asked students. "A massive asset class where most of your Harvard Business School classmates don't already have an advantage?" Change is coming to water as a financial asset class, and that volatility will provide huge opportunities for financiers who establish an early information advantage.

Understanding the current and future capital flows in water is more than an opportunity for financial gain, it is an imperative to bringing new solutions to market to solve the world's water needs. What do water capital markets look like now? What changes are coming? What would students in a real asset finance class at Harvard Business School see in five years versus today?

Let's start to answer these questions with the current view of the public sector. Catalyzing the twenty-first-century water industry will require a new formation of private investments. Private equity and venture capital often receives most attention in imagining the future of water solutions. Yet the bigger dollars lie in government funding and public debt and equity markets, and those markets have ripple effects on private equity financing. Thus, we have intentionally placed the private equity and venture capital discussion last.

Public sector – public funding and private investment

Cities often understand the investments needed to supply clean water to their customers but struggle with the political realities of water pricing and funding water projects when compared to other investments. Los Angeles, California provides an example of the challenges in funding water projects.

In Los Angeles, the second largest city in the United States, there has been a concerted effort by the water department to get its nearly 4 million residents to conserve and value their water supply. Los Angeles largely operates in a state of water deficiency and has enforced mandated water rationing. In the 1980s, Los Angeles was forced to stop using about 40 percent of its drinking water because of contamination.[2] Now the city is challenged less by contaminated water and more by quantity – Los Angeles is a coastal desert and has to import its water from hundreds of miles away.

In California, water flows southward, from the High Sierras on down to the more arid parts of the state. By the time Los Angeles gets its share of water, there likely have been myriad interruptions to the flow, quality and therefore overall supply.

A brief discussion of the complexity in supplying water to Los Angeles illustrates the importance of, and investment in, infrastructure.

Water that flows from the tap comes through pipes in buildings that snake their way into the ground. From there, piping hooks up with the main waterline.

That water line is managed by a water agency, which secures bigger pipelines that lead to reservoirs. Reservoirs are filled by rainfall and by aqueducts stretching for miles – hundreds of miles – which in turn connect to rivers, streams, and lakes. California's aqueducts connect to states as far away as Wyoming, Colorado, Utah, Arizona, Nevada, and New Mexico. The rainier north coast and upper Sacramento Valley feeds much of the aqueduct system, too, with water along with runoff from snowmelt or rainfall in other places (as previously discussed, California essentially moves water from north to south with a significant energy requirement).

One would tend to think that these engineering marvels would garner respect and attentive maintenance. After all, people tend to complain when we don't have hot water, never mind any water at all. Try going even one way without running water and respect for water delivery is ensured for life.

But that isn't the case. Water and water systems themselves are taken for granted. There is largely a lack of end-user understanding of the reality of investment in water to ensure clean and reliable supplies. Ignoring such a fundamental facet to individual and societal existence, however, is a problem.

Remember Petra? All that work, all that planning, all that engineering and infrastructure now lay in ruin. It's an ominous portent of what may occur to the water systems in modern times if innovation and upgrades aren't funded properly.

Ensuring reliable water supplies requires not just capital investment but ongoing investment and maintenance. For example, in the US, infrastructure is aging and that investment needs will continue to escalate. According to the American Society of Civil Engineers, if current trends persist, the investment required will amount to $126 billion by 2020, and the anticipated capital funding gap will be $84 billion. Moreover, by 2040, the needs for capital investment will amount to $195 billion and the funding gap will have escalated to $144 billion, unless strategies to address the gap are implemented in the intervening years to alter these trends.[3]

The public sector faces challenges to investing in water infrastructure. Although, as we've cited, expected economic returns from investing in water infrastructure are high over time, it costs money to repair and/or upgrade water systems. With a distressed economy, citizens would have to pay higher water rates and/or taxes to fund water investments. Most people would prefer to defer those actions. The current state of water in terms of availability, quality, and infrastructure, is troubling. Many water and sewer agencies can find that many of their customers are unmetered, many agencies have primitive billing systems, inhibiting their ability to collect money and data from water users, and many could dramatically reduce their operating and capital costs with fairly modest investments in proven technology to smartly manage their systems.

However, with these challenges come opportunities.

One such opportunity is how public–private partnerships can address the need for investment in the US water infrastructure (similar to alternative funding initiatives elsewhere in the world).

With public action unavailable to fund major projects, the private sector can play a role in public service on a grand scale. In recent years, the spending gap between required maintenance capital and actual spending has pushed along, maintaining the bare minimum without closing the gap. Yet, with expected high returns from those investments, private capital providers increasingly show interest in funding the gap.

US Representatives Bill Pascrell, Jr. (D-NJ) and Geoff Davis (R-KY) and Senators Robert Menendez (D-NJ) and Mike Crapo (R-ID) introduced the Sustainable Water Infrastructure Investment Act of 2011. The Act would remove state volume caps on private activity bonds (PABs) for water and wastewater projects, freeing up billions of private capital dollars for investment in the nation's water infrastructure.[4] Essentially, what the Act does is open the door for private water investments.

Jeff Sterba, the chief executive of American Water Works said in a prepared statement that:

> the legislation will play a key role in enabling investment in our nation's water infrastructure. Municipalities face competing demands for scarce financial resources right now and water infrastructure repair and improvement, though vital, is often deferred. Public-private partnerships are an essential part of the solution and this legislation will go a long way towards making more such partnerships a reality.[5]

Sterba went on regarding the IRS rules and suggested changes. The point is that to encourage capital investments in public infrastructure, there will need to be changes in basic issues such as tax rules and funding mechanisms.

Governments and the private sector can benefit enormously from such collaborative and cooperative efforts. Indeed, two highly respected scientific water experts, Bill Jury and Henry Vaux, Jr., recently issued this perspective: "Without immediate action and global cooperation, a water supply and water pollution crisis of unimaginable dimensions will confront humanity, limiting food production, drinking water access, and the survival of innumerable species on the planet."[6]

This perspective is echoed by the World Economic Forum in their 2012 report, which identified water supply risk as *high likelihood and high impact*.[7] The challenge of how to invest in public infrastructure projects is not a challenge unique to the US. It is a global risk requiring global solutions.

The urgency for investment is real yet there are significant challenges in funding water infrastructure projects.

How does a public–private water project find financing? Frequently, these are large projects, requiring tens or hundreds of millions of dollars. These projects include requests for proposals to provide a forum for competition to finance the opportunities.

Usually intense competition, and municipal tax advantaged funding, lead to low returns. Some approaches, like private activity bonds, attempt to facilitate better paths for public–private partnership.

At present, municipalities are cash-strapped. While the economic and social benefits of these projects are significant, funding such projects with current rates is a challenge (as we discussed water is not priced according to value). To complicate this further, increasing water rates is not popular or easy to do.

Moreover, the public sector in the US has seen a steady decline in funding since the 1970s and there is no comprehensive and unified federal water policy (this is further complicated by US water law, as discussed in Part II). The US is further challenged in that its funding of water innovation lags other countries on an absolute basis, on a relative basis, and has been declining. Over the last 30 years water resources research funding has decreased from 0.0156 percent to 0.0068 percent of the GDP, while the portion of the federal budget devoted to water resources research has shrunk from 0.08 percent to 0.037 percent. The per capita spending on water resources research has fallen from $3.33 in 1973 to $2.40 in 2001. This is in contrast to investment commitments from other countries such as Canada, Australia and Singapore (previously discussed in Part II).

We have each enjoyed the opportunity to speak with some of the worlds' most sophisticated investors about water. Over the past few years, many public market investors have tuned in to the water theme. They have seen the growing chasm between increasing demand and supply and the increasing value of water and sewer as commodities, and want to invest in those trends for capital gain. Yet those investors have, broadly speaking, not deployed significant capital in water investments. They frequently come to us with the same story, which sounds something like: "We get water. A few years ago, we wanted you to tell us about big themes. Now we want to invest. But how can we invest in those themes? Where are the opportunities?"

The public investment landscape offers many ways to invest in the water industry, from utilities to exchange-traded funds (ETFs) to pooled funds to multinational corporations. However, there are few "pure plays."

For example, investors lack a diversified approach to gaining exposure to up-and-coming water technology. To invest in that theme, investors would have to assemble their own private portfolios. Even investing in funds who make private investments in water companies would not offer a proxy for the increasing value of innovative solutions to water needs. There are few water-focused private investment firms, and their portfolio companies, even if one owned a small share in each one, would not create a broad enough swath to properly follow the theme big investors seek.

For publicly traded stock, people could buy ETFs, but many investors doubt that such funds truly follow the themes they seek in water. As evidence of investor interest, many entities have formed ETFs or similar vehicles. Fidelity Investments has launched a strategy called "thinking big" that explicates the need for water investments, among other world-changing ideas:

Because global water consumption is expected to increase by 40 percent over the next 20 years, water shortages may get more acute and widespread, spurring more reliance on desalination technologies, water reuse and conservation. The results could have massive economic, ecological and geopolitical consequences, creating investing opportunities in places you may never have considered.[8]

In addition to Fidelity there are other options for investments in the water industry, as outlined by David Sterman in an article titled, "Thirst For Water ETFs."[9] The water ETFs aim to own a broad spectrum of water-related companies such as makers of pumps, valves, filters, and such, making this more of an industrial angle than a utility angle. That focus has paid off: "The drought has certainly sparked interest in these types of companies," says Todd Rosenbluth, an ETF analyst at S&P Capital IQ.

A couple of the ETF's mentioned by Sterman are Invesco Power Shares Water Resources ETF (PHO), launched in December 2005 with a market cap of $807.5 million, and the Guggenheim S&P Global Water Index (CGW), with a market cap of $204.41 million (December 2012).[10]

These funds are working to create a true proxy for the real value of water. Not all funds have survived over the past few years, speaking to the challenge in valuing water and creating a viable capital market for water. The challenge they face is finding underlying stocks that mimic that value. Instead, most funds are left with investments in large, diversified conglomerates such as Veolia, Siemens, GE, and Danaher, for whom the themes of water may influence, but do not dictate their share prices.

Some entrepreneurs have developed businesses around providing capital market access to water themes. John Dickerson, decades ago, saw both social and financial impetus for water investments around the globe. Dickerson is a water investor. He runs a successful San Diego, California-based water fund, Summit Global,[11] which invests in utilities, private water companies, and big multinational water concerns. Dickerson has built one of the few asset management businesses focused on water and his funds account for a meaningful percentage of water capital markets. Clearly, Dickerson sees water as an investment opportunity.

Public debt markets

Public debt markets are critical to building the twenty-first century water industry. However, they are do not accurately reflect water issues in pricing of debt markets. Accurately factoring water risk into the municipal bond market will help capture the true value of water (see Parts I and II on water risk and the value of water).

In an article titled "Water Scarcity a Bond Risk, Study Warns," Felicity Barringer and Diana B. Henriques state that the municipal bonds that help finance a major portion of the nation's water supply may be riskier than investors realize because their credit ratings do not adequately reflect the growing risks of water shortages and legal battles over water supplies, according to a new study.[12]

As a result, investors may see their bonds drop in value when these risks become apparent, and water and electric utilities may find it more expensive to raise money to cope with supply problems, according to a Ceres report titled *The Ripple Effect*.[13]

Looking at significant water bond issuers across the southern part of the US, the report concluded that Wall Street's rating agencies had given similar ratings to utilities with secure sources of water and to those whose water sources were dwindling or were threatened by legal battles with neighboring utilities. Consequently, the study warned, "investors are blindly placing bets on which utilities are positioned to manage these growing risks."

As the study notes, the cost of tapping new supplies and repairing old infrastructure would bear down on the country's 54,000 utilities just when consumers, squeezed by a weak economy and high unemployment, were already balking at significant rate increases to cover payments to bond investors.

At the time, this Ceres report was controversial (some would argue it remains so). However, the most recent Ceres report on the topic met with apparently less controversy.

The June 2012 Ceres report, titled *Clearing the Waters: A Review of Corporate Water Risk Disclosure in SEC Filings*, expanded on the 2010 report. The key findings of the 2012 report are:

- Disclosure of water risk has increased.
- More companies are making the connection to climate change.
- Disclosure on water management systems and performance is growing but still limited.
- There is a lack of quantitative data and performance.
- There is limited discussion of supply chain risk.[14]

This is not just the case of the bond market. Increasingly, companies are reporting on water risk in response to investor interest. As previously discussed, the CDP Water Program report highlighted the water-related risks and business opportunities by the responding companies.[15]

The December 2012 Ceres report titled *Water Ripples: Expanding Risks for US Water Providers* goes further. Here are its conclusions:

- Water stress has continued to intensify.
- The market is beginning to change the way it prices water risks.
- Declining revenue and rising costs are exacerbating water supply challenges.
- Protecting future water demand is a highly uncertain proposition.[16]

Evaluating and reporting on water-related risks (through CDP Water Program and/or regulatory requirements) is a significant driver in moving from a water pricing to a water value mindset. This shift in truly valuing water will accelerate investment in water tech (more on this later in the book). The municipal bond market comprises a major share of water capital markets. If investors increasingly consider water and sewer issues in pricing and trading bonds, it could pave the road to better water management.

Private capital markets

Before we discuss private sector funding for water tech, let's clarify that water tech is not cleantech. While it is tempting to lump water tech into cleantech, the approach is flawed and water tech must distinguish itself from other technology investments such as renewable energy technologies. Cleantech is a term of convenience, used differently by different people in different contexts. The billions of dollars of investment that flowed over the past decade or so into what people often call "cleantech" actually flowed primarily to energy companies, especially in solar and biofuels. It may be convenient to group all environmental technologies together as though they represented an asset class, but the subsectors are different. Water companies rely on different drivers from other "cleantech" companies, and generally require far less capital than "cleantech" counterparts that have become well-known investment failures in the past few years.

Yet, water companies now face a common challenge. Investors are now fleeing "cleantech" investments. If water financiers do not distinguish themselves as a separate asset class, there will be ongoing limitations on private funding for water investment. The current exodus of capital from the cleantech sector contrasts with growing interest in water tech, but the sources of capital often lead back to the same investors. If those investors do not view water separately, water investment will suffer, and in turn, progress toward a 21st century water industry will stall.

As previously mentioned, there has been an exodus of capital from the cleantech sector. The exodus of capital from the cleantech sector arose from a variety of factors, including valuations on cleantech investment rounds experiencing significant downward pressure, difficulty raising additional capital for unprofitable companies with long histories but only limited customer traction, higher than expected capital requirements for some cleantech companies, and low natural gas prices from shale gas making it difficult for many energy-related cleantech firms to compete economically with their products.[17]

Water investments, unlike certain sub segments of cleantech, have benefited from healthy acquisition appetite from existing players, as well as large companies seeking entry into the water market (discussed below). A 2012 article in *Forbes* provides an overview of the activity and opportunities:

> Mergers and acquisitions of water companies jumped last year, even as venture-capital deals in water technologies fell, Cleantech Group CEO

Sheeraz Haji told me Wednesday. Global water acquisitions totaled $12.7 billion in 55 deals in 2011, up from roughly $900 million in 40 deals in 2010. Meanwhile, venture-capital investments in water dropped to $224 million in 40 deals in 2011 from $272 million in 53 deals in 2010.

The mergers-and-acquisitions number is weighted by a couple of big deals, which make it difficult to tell whether the increase signals a trend or merely a one-time spike. Minnesota-based Ecolab, which makes cleaning supplies, in July bought Nalco Holding, a waste- and water-treatment company in Illinois, for $5.4 billion. And Hong Kong infrastructure company Cheung Kong Infrastructure Holdings bought Northumbrian Water, which supplies water and wastewater services, for $3.9 billion. Those two large deals made up 73 percent of last year's total, meaning that most of the deals were much smaller.

... a few trends are driving the water buys. Before now, there weren't been [sic] many mid-sized companies in water for these giants to buy – and there still aren't many – but, as some small companies grow into the $15 million and $20 million range and beyond, they are getting snapped up, Haji said. "If you grow your business up to $20 million and you have a growth path, life is good," he said.[18]

But what about earlier-stage businesses? And how do these water tech companies secure private funding? The following chapter discusses how organizations (such as Imagine H2O) can support funding and commercialization.

Most water businesses are bootstrapped, or rely on commercial bank loans, or friends and family investments. The prototypical small water tech company originates from the mind of an engineer. Often, that engineer worked for a city or utility, or otherwise experienced water problems first hand and dreamed of a smart method of solving those problems. His or her first customers are frequently local. The engineer, leveraging local relationships and a smart solution, may reach profitability on a small scale, but would need additional capital investment to scale the business. This story illustrates the roots of fragmentation in the water industry, and paints a picture of an industry of brilliant tinkerers.

How, then, can we provide capital to achieve our vision for a twenty-first century water industry? How can early stage water tech companies cross the chasm into mainstream adoption?

These questions, and working hand-in-hand with talented founders of small water-tech companies, led to the development of Banyan Water. I (Pechet) believed that we could accelerate the adoption of smart water management faster by scaling up existing, working solutions than developing new ones from scratch. Banyan Water's strategy relied on raising institutional private capital, and assembling a team to help take existing, proven technology and bring it to market more broadly.

The strategy is far from novel, but could provide a bridge to institutional capital for some water tech entrepreneurs who otherwise might not have access to the money and resources needed to grow to significant scale. What are some of those capital resources?

Socially responsible investing and impact investing

Over the past several years we have seen the emergence of socially responsible investing (SRI) and impact investing. Both forms of investing can provide options for water tech companies.

SRI, also known as sustainable, socially conscious, "green" or ethical investing, is any investment strategy, which seeks to consider both financial return and social return. In general, socially responsible investors encourage corporate practices that promote environmental stewardship, consumer protection, human rights, and diversity. Some avoid businesses involved in alcohol, tobacco, gambling, pornography, weapons, and/or the weapons. The areas of concern recognized by the SRI industry can be summarized as environment, social justice, and corporate governance – as in environmental, social governance (ESG) issues. In addition to stock ownership either directly or through mutual funds, other key aspects of SRI include shareholder advocacy and community investing.

The relative social impact matters because it a key aspect of one of the fastest growing segments of the financial industry, something called "impact investing." Impact investing combines doing good with getting a good return on your investment.

Some of the wealthiest people, biggest foundations, and most prestigious financial institutions have turned to impact investing to replace portions of their philanthropy and/or their venture capital investments.

One of the founders of eBay, Pierre Omidyar, is investing in start-ups this way. So are Bill Gates, The Rockefeller Foundation, JP Morgan and a slew of other sophisticated investors.[19]

Impact investments are investments made into companies, organizations, and funds with the intention to generate measurable social and environmental impact alongside a financial return. Impact investments can be made in both emerging and developed markets, and target a range of returns from below market-to-market rate, depending upon the circumstances. Impact investors actively seek to place capital in businesses and funds that can harness the positive power of enterprise.

In addition, impact investing occurs across asset classes, for example private equity / venture capital, debt, and fixed income. Impact investors are primarily distinguished by their intention to address social and environmental challenges through their deployment of capital.

Venture capital, angel investors, and private companies

Venture capital firms (VCs) or angel investors are the most common capital providers for early stage water tech businesses. Despite the aforementioned tiny fraction of innovation capital flows into water over the past few years, VCs and angels have climbed the education curve on the water industry. That is a necessary precursor to more investments.

The water sector is poised for increased investiture. "Technology investment seems to be one of the fastest growing areas of interest in the water sector at the moment," GWI notes, going on to explain:

> What seems to have happened is that the venture capital sector had to reinvent itself in the wake of the dot.com bust of the early 2000s, and many funds took up the greentech theme. There was a rush to invest in renewable energy start-ups, which quickly became over-valued. Green tech funds then started to look at water, but few have actually invested. The most important investors in the sector remain the old hands: XPV Capital from Toronto, FourWinds Capital Management in London, Emerald Technology Ventures in Zurich (which recently invested in leakage software specialist TaKaDu), Arison and Israel Cleantech Ventures based in Tel Aviv, and Kinrot Ventures based in northern Israel.
>
> There are two main reasons why there are a lot more greentech funds looking at the water sector than actually investing. The first is that the fragmentation of the water business means that the scope of individual technologies appears limited. For example the biggest water technology of the past 50 years is probably the reverse osmosis membrane – but RO membranes are not even a $1 billion market. Compare that to renewable energy companies which have their eyes on a market currently valued in the trillions of dollars a year, and you can understand why water is less exciting close up.
>
> The second problem is the long adoption cycle in the water sector: most water technology start-ups take at least five years to go from proof-of-concept to first commercial reference. This is because the end-user customers – typically public water utilities – have no competitive or profit-seeking incentive to take risks on new technologies. Even if they want to take the risks, few of them have the money to spend. The established venture capital investors – the XPVs and the like – know their way around these problems, and can make good returns on smart investments. Most of the newcomers can find these problems challenging.
>
> Fortunately there are some exceptions. Three of Silicon Valley's notable VCs have taken positions in water technology. Khosla Ventures has invested in NanoH20 and Calera; Kleiner Perkins has invested in APT (along with XPV and others), and Draper Fisher Jurvetson has invested in Oasys. They have a much bolder style of investment than the established water technology investors. They are prepared to take bigger risks up front in order to accelerate the rollout of new technologies. They need to succeed. If NanoH20, Oasys and APT can force through a route to market in less than five years, they will have established a way through water's tortuous procurement process that others can follow.[20]

The good news for innovators and entrepreneurs is that the backlog of investments cannot continue. As GWI admits, "With the challenges we face, this cannot go on."

One of the more notable VCs is Vinod Khosla,[21] and he is paying attention to water tech having invested in companies such as Driptech and NanoH20. In 2012 Khosla Impact Equity, a new social venture fund, invested in Driptech,[22] a for-profit social enterprise that makes affordable irrigation products for small-plot farmers (refer to Part II on opportunities in smart/precision agriculture). In addition, Khosla invested in NanoH20, a start-up developing technology that can clean water with low amounts of energy (refer to Part II on the energy–water nexus).

NanoH20 (see Box 8.1) has been working on nanoengineered reverse osmosis membranes for desalinating water that are supposed to be more productive and use less energy than traditional desalination membranes. The problem with standard reverse osmosis is that it is energy-intensive, and that makes it costly. NanoH20 says its membrane is much more permeable than prior attempts, which could translate to less energy needed to push the water through and lower facility-operation costs.

In addition to VCs there are "angel investors." Angel investors can be friends and family members. More and more angel investors are professional investors who act as the first step on the VC ladder. Promising for water tech businesses is the fact that there are angel investor networks aimed at solving social and environmental problems.

Investors Circle (IC)[23] is one of these networks, and Suzanne Beigel, its London-based chief executive, says water is a growing investment space that angel investors are excited about. IC holds several venture fairs per year, both virtual and in-person events where entrepreneurs pitch their ideas to investors. Other angel networks, such as the Social Venture Network and Slow Money, among others, work in a similar manner.

Another important source of private capital is strategic investment by large companies in the water industry into emerging companies and technologies. Strategic investment is especially valuable in the water market due to the challenges of reaching customers and gaining credibility as a proven vendor to those customers. A well-known, large industry player can be an ideal go-to-market channel for an emerging water tech company. Companies such as BASF, GE, Veolia, and Siemens have invested actively in water tech companies, pairing financial capital with strategic partnership to accelerate growth of those investments. Veolia has announced a significant effort to create partnerships in Silicon Valley and beyond, using Veolia's customer base and sales capabilities to help emerging water tech companies reach customers. Many venture capitalists and angel investors prefer water tech investments with a strategic industry co-investor for those reasons.

Funding a water tech company and operating a water tech company are two very different things. The commercialization of water technology products takes a very unique set of management, operating skills, and foresight to become successful.

Box 8.1 Case study: NanoH20

NanoH20 is an interesting example of how water industry outsiders are changing the water industry.

Jeff Green is an internet entrepreneur who decided water tech was a promising opportunity. According to the NanoH20 website:

> Jeff Green founded two software startups, Archive, Inc. (sold to Cyclone Commerce) and Stamps.com, Inc. (NASDAQ: "STMP"). In both companies, Mr Green co-wrote the original business plan and played a key role in raising over $300 million in private equity, public and debt financing. At both companies, Mr Green was responsible for directing the company's strategy, business model and marketing operations, including product management.[24]

NanoH20 was spun out of research based on technology developed at the University of California Los Angeles (UCLA). NanoH20 develops membranes for water desalinization. The El Segundo-based start-up attracted venture capital from Kholsa Ventures and Oak Investment Partners. It went on to win an Aquatech Innovation Award, and received yet another round of funding. Because it designs, develops, manufactures and markets reverse osmosis (RO) membranes that change the fundamental economics of desalination, the company helps bottom lines. The membranes dramatically improve desalination energy efficiency and productivity – flipping the famous downside of desalinization to produce strong upside.

The technology has received several industry awards, which goes to show the value of competitions: it attracted even more capital after winning!

Notes

1   Professor Perold serves as a member of the advisory board of both Banyan Water and Imagine H2O.
2   www.smarterflush.com/in-the-news.php.
3   American Society of Civil Engineers (2011) *Failure to Act: The Economic Impact of Current Investment Trends in Water and Wastewater Infrastructure*, Reston, VA: ASCE.
4   www.wateronline.com/doc.mvc/American-Water-Supports-Congressional-Action-0001.
5   Ibid.
6   Quoted in Thomas M. Kostigen (2008) *You Are Here: Exposing the Vital Link Between What We Do and What That Does to Our Planet*, New York, NY: HarperOne, pages 168–9.
7   World Economic Forum (2012) *Global Risks Landscape*, Geneva: WEF.
8   http://thinkingbig.fidelity.com.
9   David Sterman (2012) "Thirst For Water ETFs," *Financial Advisor*, 14 December, www.fa-mag.com/news/thirst-for-water-etfs-12857.html.
10 http://seekingalpha.com/symbol/pho.
11 www.summitglobal.com.
12 Felicity Barringer and Diana B. Henriques (2010) "Water Scarcity a Bond Risk, Study Warns," *New York Times*, 20 October.

13   Ceres (2010) *The Ripple Effect: Water Risk in the Municipal Bond Market*, Boston, MA: Ceres.
14   Ceres (2012) *Clearing the Waters: A Review of Corporate Water Risk Disclosure in SEC Filings*, Boston, MA: Ceres.
15   www.cdproject.org.
16   Ceres (2012) *Water Ripples: Expanding Risks for US Water Providers*, Boston, MA: Ceres.
17   Walter Frick (2013) "There's No Denying It: The Cleantech VC Exodus is Truly Here," Cleantech Group, 4 January.
18   Jennifer Kho (2012) "Water Acquisitions Rise: Will Venture Capital Follow?" *Forbes Green Tech*, 28 February, www.forbes.com/sites/jenniferkho/2012/02/28/water-acquisitions-rise-will-venture-capital-follow.
19   For a longer list of potential funders and a tutorial, see www.thegiin.org.
20   www.globalwaterintel.com.
21   www.khoslaventures.com.
22   www.driptech.com.
23   www.investorscircle.net.
24   www.nanoh2o.com.

Chapter 9

Commercialization

Commercialization is perhaps water tech's greatest challenge and, in turn, where change can have the greatest impact. While the gravity, breadth, and variance among water problems requires a greater supply of innovation, many great ideas currently lie dormant. The industry of tinkerers has developed countless solutions that would improve water and sanitation management, but those ideas have yet to reach mainstream adoption. One can imagine a new treatment technology sitting on a garage shelf somewhere. Garage entrepreneurs certainly exist in water. The great near-term opportunity lies, however, in taking exceptional ideas that a few customers have already adopted, on a local level or in fragmented fashion, showcasing these, and providing a pathway to broad adoption.

Early in the innovation and funding process a critical question must be asked and answered: can this technology be commercialized and, if so, how? What is the path forward? What are the resources needed (management team in particular)? How quickly can we scale the business to ensure long-term profitability and financial success for investors? Although that may seem obvious, in an industry of tinkerers, many of whom invent novel improvements to the status quo of water management, far too many believe that "if we build it, customers will come." For an innovator working to improve water treatment, for example, who looks at the need for better water quality and understands the impact he or she can have on the world, the opportunity may seem so clear that it's easy to overlook the buying process.

Any new technology faces challenges in the commercialization phase, but the water industry's commercialization challenges present special obstacles.

We have identified the challenges unique to water in previous chapters, but to recap:

- *Disconnect between the price and value of water.* Water is essentially free and, as a result water, technologies are saddled with long payback periods (if one just evaluates payback based upon the current price of water).
- *Water tech is not just about technology.* Water has economic, environmental, social and cultural dimensions all of which must be concurrently managed (unlike resource issues such as energy). There are values associated with water and as a result reputational risk and brand value is important.

Stakeholders care about how water is used by the public and private sectors. For example, these stakeholders can impact social license to operate for private sector companies and perceptions of water reuse and advanced metering systems for public sector water utilities.

- *Wide variation in legal ownership and regulatory frameworks for water depending on political setting.* The US is a good example, with regional and state differences in water law coupled with numerous regulatory agencies responsible for water.
- *Lack of funding for water technologies and infrastructure in the public sector* (ongoing budget cuts and pushback on government funding results in an aging and unreliable infrastructure).
- *A fragmented marketplace for water and how it is managed.* Numerous water supply and treatment utilities, along with different "types" of water (fresh, potable, brackish, salt, gray water, etc.).
- *Risk-averse culture in the water industry.* Risk taking is not readily embraced in part due to the need to safeguard public health in managing water supplies. This also drives the need to pilot technologies to "guarantee performance." Customers require long testing periods with lots of proof points and pilot tests.
- *The path to commercialization is not clear-cut.* The sales distribution channels for water tech are difficult to understand and to penetrate.

There is a fundamental lack of knowledge about water issues, including valuations, technology advancements and solutions. Therefore going from innovation to commercialization, even with proper funding, can be difficult.

The lack of knowledge of water issues creates an opportunity. "The key thing is to focus on making customer propositions which are about taking costs out and improving efficiency. Water has an interesting role to play in this. It is both part of the problem and part of the solution," says Christopher Gasson, the publisher of *Global Water Intelligence* magazine.[1] For many water innovations, the key customer value proposition has little to do with water, but instead relies on the effect water can have on other processes, such as saving energy, improving agricultural crop yield, protecting property from water damage, managing reputational risk and many others.

In an article titled "Opportunities in the Malaise," Gasson says that the water industry has changed:

> From now on it is going to be about delivering efficiency and cutting costs. Rising water costs are part of the trend towards increased raw materials costs. As with copper, oil, gas, and other natural resources, the easy reserves have been tapped. The marginal cost of tapping new water resources has been rising for some time. If the water industry can provide the technologies that enable customers to use water more efficiently, it will thrive. Furthermore, if water technologies can help reduce the cost of extracting other primary products such as fossil fuels, minerals and agricultural products, it could grow

strongly against a bleak overall economic trend. We are already seeing this happening in the way natural resources companies are accelerating their investment in water technologies.[2]

Despite these unique challenges in commercializing innovative water technologies there are an increasing number of success stories of how innovative water technologies have been commercialized, even to municipal and utility water customers. Success stories are crucial to greasing the wheels of water tech commercialization. Every investor wants to hear success stories and profitable exits before investing in water tech. Every innovator uses those successes as fuel to burn through difficult times in the commercialization process. Even customers need to gain comfort with early adoption of water tech, and looking at other customers who benefited from taking risks on water tech helps.

Tom Pokorsky offers one example of successfully commercializing new innovation to municipal customers. Pokorsky has a unique view and perspective on funding in the water tech sector. As president and chief executive of Aquarius Technologies, an advanced wastewater treatment company in Port Washington, Wisconsin, Pokorsky gives the view from running a start-up. His more than 30 years in the water business also adds a dimension of history and context to where the sector is now.[3]

Pokorsky started in water business in 1979. This was at a time when the Clean Water Act of 1972 was creating all sorts of deals and jobs for the water industry. Municipalities had to live up to new standards and that meant new technologies were needed.[4]

At that time, Pokorsky worked for a wastewater treatment company, Sanitaire, whose owner was retiring. Pokorsky and three other executives led a leveraged buy of the firm. "We got a bank to give us a loan based on the company's assets and that we could pay back out of future profits," Pokorsky recalls, adding that there is no one who could get that type of loan now. Even then (in the early 1980s) he said that he had to use his house as collateral: "I was young so it wasn't like I had any equity in it. The bank just wanted the threat of being able to take it away." It turned out that that would be far from the case.

Pokorsky timed the buyout well. The mid- and late 1980s were the wonder years in terms of financings – leveraged buyouts, junk bonds, and other mechanisms rained riches on Wall Street and financiers like never before. New tax laws put more money in people's pockets. Moreover, there was still a lot of work to be done in both the safe drinking water and wastewater segments of the business.

By the 1990s, Fortune 500 companies were eyeing water companies as acquisition candidates. Pokorky's Sanitaire had been built up to a $40 million a year business. With strong growth and cash flow, the fundamentals were attractive to big companies' balance sheets. Indeed, in 1999, ITT purchased Sanitaire. They made Pokorsky head of their new wastewater division, and gave him money to spend to grow the business. "It was a dream come true," he recalls. "I had a Fortune 500 company telling me to spend money to grow the business." And he did, building a $350 million division (from $40 million platform) by 2006. Fifty percent of the division was built on growth, and fifty percent on acquisitions.

"In the era between the mid 1970s and the mid 1990s anybody with a decent idea could get a good jump start because there was so much work [getting water to adhere to new standards]," Pokorksy says. The Safe Water Drinking Act was originally passed by Congress in 1974 to protect public health by regulating the nation's public drinking water supply. The law was amended in 1986 and 1996 and requires many actions to protect drinking water and its sources: rivers, lakes, reservoirs, springs, and groundwater wells. The Clean Water Act was significantly reorganized and expanded in 1972 and amended in 1977. These two laws required billions of dollars in funding and made new technologies exigent.[5]

From about 1995 to 2005, Pokorsky says, was the era of the Fortune 500 companies because they gobbled up any business with revenue between $15 million and $100 million. Then everything changed. Financings were getting out of hand. Acquirers were paying 20 to 30 times EBIT (earnings before income and taxes). In some cases, startups were getting bought for multiples of their sales numbers because they had little or no profit. Although buyers commonly acquire early or pre-revenue companies based on sales multiples in many industries, the acquisitions in the water industry indicated the feverish growth expectations among acquirers. General Electric (GE), for example, paid $656 million in 2006 for Zenon Environmental, a company with approximately $200 million in revenue that lost money the prior year.[6]

GE paid a price based in part on Zenon's advanced filtration water tech. The water industry, long thought of as a slow-moving, industrial and municipal business, had become hot.

According to Pokorsky, "over-the-top valuations caught the attention of the heavy gamblers of the financial industry, venture capitalists." Their strategy, according to Pokorsky, was to seed entrepreneurs to develop new technology, and then sell it off at a high multiple.

Pokorsky had retired around this time but venture funds started calling him. They offered to back him in a new wastewater technology business. He took the bait.

What happened next speaks to the business of funding water technology today: the downturn in the global economy hit the entire capital supply chain like a tornado.

"Over the past 30 years I have never seen a downturn in the economy affect the water business. It has this time," Pokorsky says, "because municipalities are broke. They can't afford to make payments. They are delaying projects."

In spite of those challenges, Pokorsky has once again commercialized a novel water technology. His new business uses bacteria to digest wastewater sludge, and then, in a staged process, digest itself. The process promises to reduce wastewater operating costs and landfill use. Although Pokorsky has faced the challenge of municipal funding shortages in commercializing his new company's water tech, he has nonetheless built a growing and profitable business changing the way water is managed.

How to promote commercialization?

How do we overcome the challenges of commercializing innovative water technologies? One of the opportunities to catalyze water tech lies in borrowing from successful commercialization pathways in other industries, and promoting them for the water industry. A logical starting point are academic institutions that successfully offer research and development platforms in other industries.

The university technology transfer office (TTO) has emerged as a viable platform to identify and support the commercialization of water technologies, and to provide entry into the many water-related activities within a university. The progress of university TTO offices over the past several years bodes well for the future of water tech.

By definition, the TTO is:

> the process of skill transferring, knowledge, technologies, methods of manufacturing, samples of manufacturing and facilities among governments or universities and other institutions to ensure that scientific and technological developments are accessible to a wider range of users who can then further develop and exploit the technology into new products, processes, applications, materials or services.[7]

University TTOs can perform a wide range of services related to patenting and licensing of inventions, establishing research partnerships with industry, and in negotiating the exchange of research materials and tools. Universities worldwide have recognized that economic development is "a legitimate purpose" of higher education.[8]

Figure 9.1 shows the role TTOs play in the development of a new technology.

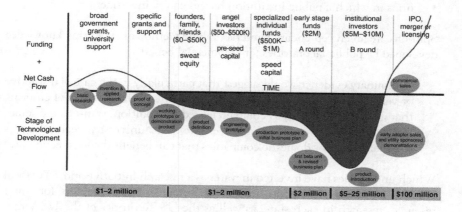

Figure 9.1 Conceptualization of the role TTOs play in developing new technology[9]

TTOs have spawned numerous water tech successes. Earlier in the book we described how John Schroeder joined a university-backed TTO start-up and took it through to acquisition by Graver Technologies. This is an excellent example of how water tech can get its start. Yet, until the past few years, most university TTOs have not focused on water.

In 2006, I (Pechet) went to several top US universities to find promising water technologies. I found little activity, and frequently received the question "Do you mean energy technology?" Now many of those same universities have partnered with Imagine H2O and other water organizations, offered water courses, and shown a new commitment to water tech. The engagement of academic institutions in next generation water tech will create new industry leaders and new centers of research and support for innovation.

Why are TTOs particularly important to the next-generation water tech initiative? They are proven pathways to capital formation around new technology, thereby offering support in overcoming the challenge of raising early-stage capital for water tech.

Linking TTO development with a specific government grant, as suggested by Figure 9.1, could replace an angel investment round. University partnership in technology development can often improve the chances of winning government grant funding. Moreover, strategic industry players and early stage capital providers frequently peer into the TTOs of top universities for first looks at promising emerging innovation. Water initiatives within TTOs are likely to gain visibility with key players in the commercialization process.

How would this work? Partnerships with the private sector can result in university "spin-offs or ventures" and include:

- firms founded by public sector researchers, including staff, professors and post-doctorate students;
- start-ups with licensed public sector technologies; and
- firms in which a public institution has an equity investment.[10]

Spin-offs are an entrepreneurial and risk-taking method of exploiting knowledge developed by public laboratories for commercial benefit:

> Preliminary evidence would suggest that particularly in the case of disruptive or very innovative technologies spin-offs might be the best way of ensuring that a given technology is commercialized. In addition, spin-offs are sometimes used as a mechanism to ensure that university technology is commercialized by domestic companies that can benefit the local economy.[11]

Which universities have invested in water as a research initiative and TTO focal point? The National Institute for Water Resources[12] links centers for water resources research in each state, as well as the US Territories of the US Virgin Islands, Puerto Rico, and Guam, for a total of 54 institutes. Although those centers do not explicitly focus on commercialization of technology, the network can provide fertile ground for novel research and methodology that lay the foundation of water tech ventures. A small sample of schools with active water tech initiatives includes the Universities of California at Los Angeles and Berkeley. UCLA, as previously noted, spun out Nano H2O. UC Berkeley's Water Center has advanced innovative thinking on water policy and the university has fostered several recent startups. MIT has a Center for Clean Water and Clean Energy, and has increased its focus on water in recent years. A group of researchers and a venture capitalist recently spun out MIT technology for novel membranes for water treatment, creating a company called Clean Membranes which has received subsequent venture capital financing. Harvard began a more concerted effort to foster knowledge and solutions for water by launching a Water Security Initiative, with the goal of collaboration across Harvard's schools and beyond to create water-secure countries. More university water centers launch annually, providing regional home bases for water tech commercialization.

Academic institutions are crucial to water tech commercialization because of their ability to link key stakeholders to the commercialization effort. Other initiatives that can link those stakeholders include regional water innovation clusters and water-focused innovation centers.

Innovation centers form partnerships between universities, business, and sometimes government. Their goal is to bridge the gap between university research and the ability of business to bring that technology to market. For water tech, finding existing efforts to bridge the innovation-to-commercialization gap can mean the difference between success and failure.

The Boulder Innovation Center,[13] for example, follows a mission to connect talent with opportunity. It affiliates with local universities and develops pathways to help new investment get to market more quickly. It also recruits top tier business talent to provide entrepreneurs with guidance.

As we have noted, the unique nuances of water tech require special, dedicated efforts. Water clusters attempt to unite a wide range of stakeholders, often including local innovation centers, as well as business, academic and government entities. These stakeholders come together to spur economic development through the sharing of ideas and common resources beyond growing an individual business.[14]

The Colorado Water Innovation Cluster[15] is an example of a regional innovation cluster dedicated to water. It was created to "serve as a catalyst and focus on innovative and entrepreneurial ways to grow the water resource and technology sector of our economy through actionable initiatives and showcase projects." According to the Colorado Water Innovation Cluster, their vision is:

> to leverage the abilities of our members to produce long-term solutions to global water issues. In the next five years, our initiatives will establish our region as a global leader in water innovation, increase regional water-related technology commercialization, and contribute to the economic vitality of our community.[16]

As an example of the interweaving of organizations to spur water tech innovation, The Water Innovation Network (WIN)[17] is a collaboration between the Colorado Water Innovation Cluster and Colorado State University (CSU). The goal is for these partners to create a platform to "foster the development of innovative solutions and cost-effective technologies focused on meeting the ever-increasing global demand for clean water." WIN aims to leverage its cluster to drive innovation within the water industry supply chain and to encourage end-users of water to adopt innovative solutions that reduce water demand and effluent.

WIN's stated objectives are to:

- stimulate innovation in the water industry supply chain to meet industry demand;
- create a collaborative network of water sector companies, organizations and universities;
- increase deal flow within the water sector;
- leverage additional national or European collaborative R&D funding;
- develop new overseas markets for the water industry supply chain; and
- increase government support of innovation in the sector and investment in R&D.[18]

Ultimately, WIN will provide a platform for academic research and private product development "to address the water challenges of the future, including such topics as homeland security and early warning systems, drinking water safety, climate change adaptation, surface/groundwater interactions, watershed planning and management, the nexus of water and energy, etc." According to WIN:

> each of these topic areas present opportunities to advance existing technology, develop new technology, and create new methods of analysis and management. This presents the opportunity for economic vitality in our community, which already possesses a wealth of expertise in the water management, treatment, and analytics fields.[19]

While these Colorado initiatives demonstrate the opportunity to support emerging water tech at a regional level, water tech will benefit from emerging efforts to link regional innovation clusters and centers.

One example is the "Center of Advanced Materials for the Purification of Water with Systems (WaterCAMPWS)," which is a National Science Foundation Science and Technology Center.[20] The mission of The WaterCAMPWS is to "develop revolutionary new materials and systems for safely and economically purifying water for human use, while simultaneously developing the diverse human resources needed to exploit the research advances and the knowledge base created." Although the organization focuses on purification, it serves as an example of a hub-and-spoke model of linking key players to support water tech.

In addition to these cooperative efforts, promoting the commercialization of water tech requires market-based, capacity-building for innovation in water. Market-based solutions have proven successful in other industries.

Boot camps and incubators have taken hold in other areas of tech, and emerging efforts to link existing organizations to water and to borrow from best practices in capacity-building by water organizations have already borne fruit.

"Boot camps" are short-term workshops to help entrepreneurs get started with a new business idea. These workshops can help the entrepreneur see the opportunities to build a business in water, and to connect with key players to help in the early stages of innovation. Water-focused boot camps offer the promise of drawing more innovators to water tech.

Innovation incubators or accelerators help entrepreneurs take existing ideas to the next stage of development, and generally include a variety of business support resources and services. Business incubators are designed to bolster and spur into action successful entrepreneurial companies. They are sometimes mission-specific and offer everything from office space to strategic advice. Local governments sometimes back incubators to spur economic activity and attract new businesses. Venture capital firms often incubate startups to creep into fully funding ventures. Increasingly, universities are also developing incubator programs.

Y Combinator and Tech Stars offer examples of incubators where innovators across a spectrum of industries apply to gain entrance into programs to help bring their businesses along the development cycle. Like many incubator programs, they offer resources in exchange for minority ownership in their incubated companies. Y Combinator and Tech Stars claim numerous successful startups, and have become focal points for early stage investors and acquirers of early stage businesses.

Similar organizations have made positive impacts in cleantech. One example, the Cleantech Open,[21] offers a competition platform to identify and support the commercialization of new technologies in clean tech industries, including water. The mission of the Cleantech Open is to "find, fund and foster entrepreneurs with big ideas that address today's most urgent energy, environmental and economic challenges." Of the nearly 600 companies the Cleantech Open has worked with about 80 percent remain viable and have gone on to raise external capital exceeding $660 million. The Cleantech Open has awarded over $5 million in cash and services (www.cleantechopen.org). The Cleantech Open began as a volunteer effort in Silicon Valley, and its success enabled hiring a full-time staff, and spreading regional clusters from San Francisco to other technology hotbeds in the US.

Interestingly, the Cleantech Open struggled with water tech, even where it succeeded broadly in its early years. The organization attracted few innovators in water compared to other clean tech clusters, found fewer promising business concepts, and fewer ready capital providers to bolster water innovators' chances of success. In part due to these challenges, key members of the Cleantech Open were instrumental in the early years of building Imagine H2O's effort to singularly focus on water and address the impediments to water tech innovation and commercialization.

Rebeca Hwang, a member of the board of directors of the Cleantech Open, joined as an early advisory board member of Imagine H2O:

> Overcoming barriers to water innovation requires gathering and motivating experts in various fields of water, customers and go-to-market partners for water tech, and educating and linking potential and existing innovators, industry players, and financiers. Water innovation has moved slower than it otherwise might because it lacked a platform specifically dedicated to these efforts.[22]

Imagine H2O has borrowed from successful capacity-building models, and linked its program with a hub-and-spoke strategy to existing innovation centers and clusters. With the vast and diverse set of challenges and opportunities in water tech, other organizations can and will apply similar strategies to boost water tech toward faster creation and adoption.

Promoting commercialization of water tech, as we have noted throughout, requires aligning policy and funding with innovation. As discussed in Part II of this book, several countries are moving aggressively to draft and implement policy measures to support water tech and funding water tech hubs.

China and India, facing challenges to growth from water and water infrastructure problems, and seeing the opportunity for leadership in water tech, are leveraging public policy to promote water innovation. China and India's multi-billion dollar investments in better water management systems offer hope for acceleration of water tech on massive scale, and highlight the interconnections of water systems across political borders (trans-boundary water).

For example, over the next decade, China plans to double the average amount spent on water conservation annually.[23] These investments include a series of canals connecting the water-rich Yangtze River of the south to the water-deprived Yellow River of the north, and desalination and water reclamation projects. "While projects like the Yangtze River-Yellow River canals are ambitious, they are not the future," points out the investment newsletter Seeking Alpha. "Merely moving water around China will not be able to supply her with the water she needs in the future, the future for water in China is water reclamation, water treatment, and desalinization."[24]

India's investment in water tech offers the promise of catalyzing new innovation within and beyond its borders.

Addressing a business forum in Singapore, India's Water Resources Ministry Secretary, Umesh Narayan Panjiar, underscored the investment opportunities in the Indian water sector, where he said $50 billion worth of investments have been made over the past five years, more than double the $22 billion amount projected for 2002–7. "The challenges faced by India in developing water sectors are the new opportunities for business," Panjiar says. He points out that water storage sites have been identified for harvesting rainwater while desalination plants would be required for the coastal areas and islands to ensure drinking water supply: "There are a number of opportunities in India for...water companies to invest, especially through the public-private partnerships."[25]

Add south Asia's population of 1.5 billion, which is growing by 1.7 percent per year, and Asia's importance to the future of water tech grows even clearer. Already, while India and China support development of internal water tech, they are importers of water tech from abroad. Late in 2012, India and Australia announced a bolstering of their water science and technology partnership, with $12 million in additional investment over a four-year period.[26]

The cross-country partnership aspect of water tech investment is a growing theme that bodes well for hopes for advanced water management globally. In 2012, the Finance Minister of Israel, Yuval Steinitz, met with Chinese Finance Minister Xie Xuren, and penciled an agreement calling for China to purchase $300 million in water technology from Israel, focused on irrigation technology to aid China's agricultural development.[27] Deals like this one can promote the commercialization of water tech as entrepreneurs and governments see success stories. Importantly, a deal like this one demonstrates that water tech can transcend a "not-invented-here" syndrome that often prevents technology adoption. These examples may further encourage policy support and coordination between government, academia, non-profit, and innovation segments of economies to promote water tech. As water tech producers seek motivated markets for accelerated adoption of their products and services, regions with policy and spending mandates can be drivers of mass adoption.

Private sector companies and innovation

Customer purchasing offers the true key to commercialization of water tech. Government, academic, non-profit, and investor finance can carry innovation only so far. Private sector companies are ramping up interest in water tech innovation for several reasons. One of the drivers is that adoption of innovative approaches to water tech is part of a movement by companies to build water stewardship strategies.

Numerous stakeholders predict increased private sector spending on water tech. Ceres, the NGO focused on water and climate issues that we discussed in Chapter 8, alerts stakeholders to the growing role for private sector purchasing in addressing water-related issues and driving technology innovation. Ceres notes that the record droughts in China and Russia, the unprecedented floods in Australia, and the ongoing legal battles over water supplies across the United States are creating major new challenges for generating power, producing food and supplying clean drinking water to growing populations. Declining water quality from industry and agriculture and inefficient water management practices are problems as well. As we are seeing, these can be opportunities for commercial applications of water tech solutions.

Consider this: according to Ceres, food and fiber production represents at least three-quarters of global water use. In the US, the electric power sector alone accounts for about 41 percent of freshwater withdrawals. "Businesses need to understand the impacts their operations and supply chains have on water supplies and put management systems in place to reduce pollution and improve water efficiency," Ceres says. "Increasingly, investors are boosting scrutiny of water risks in their portfolios and calling for better transparency and stronger action from companies to mitigate these risks."[28]

At the level of discussion and planning, many large customers agree that water tech purchasing adoption will rise. The combined perspective of several multinational companies and NGOs as to the role of water tech in addressing the projected gap between supply and demand was outlined in a 2030 WRG report.[29] The report cited innovation in water technology – in everything from supply (such as desalination) to industrial efficiency (such as more efficient water reuse) to agricultural technologies (such as crop protect ion and irrigation controls) – as playing a major role in closing the supply–demand gap through 2030.

The report specifically addressed commercialization of technologies:

> Many of the solutions on the cost curves developed for each country imply the scale-up of existing technologies, requiring expanded production on the part of technology providers. The cost curves provide a framework that technology providers can use to benchmark their products and services for an estimate of their market potential and cost competitiveness with alternative solutions. Membrane technology, for example, is still about 2 to 3 times more expensive in China than traditional treatment technologies. As the need for high-quality water treatment increases, specifically for potable or high-quality industrial use or re-use, low-pressure membrane technology could develop a market potential of up to 85 billion m³ by 2030, 56 times its volume in 2005.[30]

This is the world in which we work, and both of us – Sarni in working with companies in developing water stewardship strategies, and Pechet with Banyan Water customers – are seeing increased action in evaluating innovative water technologies to address water scarcity, water quality, protect ecosystems and provide clean water to communities in which companies operate. "Collective action" by multinationals coupled with focus on innovative water technologies is driving change in the water industry. Yet, even among the Fortune 500 companies with which we work, water tech varies in priority for purchasing, stage of understanding, and adoption.

The stage of action on water tech purchasing among multinationals and smaller companies varies dramatically. Companies in sectors in which water is a mission-critical process, such as food and beverage production, natural resource extraction, power, pharmaceuticals, and high tech manufacturing, have historically driven water tech adoption. Their requirements for process improvements, license to operate and the management of reputational risk continue to advance demand for water tech.

Lux Research, which provides strategic advice and on-going intelligence for emerging technologies such as water tech, highlights the asymmetry in likely near-term adoption of water tech among private customers. They note that it is near impossible for innovative new technologies to break into a market dominated by age-old, cheap, commodity solutions, and so it goes for the water treatment industry. Yet, emerging market drivers – from tighter water quality regulations to corporate emphasis on water conservation to water-intensive applications in oil and gas exploration – have all opened the door to new chemical treatments and non-chemical alternatives.

This is where the commercialization of water tech products gets complicated and where consultants can prove their worth. For example, Brent Giles, a senior researcher at Lux (and the lead author of the report *Water Chemicals and Competitors: The Long, Long March of the "Chemical-Free" Revolution*), says:

> Opportunities await the new wave of reduced- and non-chemical water treatments, but those opportunities are distributed unevenly across application markets. New approaches for treating municipal water, for example, won't budge conventional chemical-based methods. But in the oil and gas industry, non-chemical treatments could move very fast because their relatively small footprint enables produced water to be treated at the drill site and reused.[31]

Water tech adoption could move faster in industries like oil and gas because water is mission-critical. One executive of an oil and gas company shared with me (Pechet) that his team had evaluated over 700 different technologies and methods for handling produced water from drilling oil and gas wells. In a discussion about water and wastewater megatrends, Hubert Fleming, former global director at Hatch Water, identified water demand in the natural resource space (oil and gas, mining, agriculture, etc.) as one of the key drivers that will advance water tech.

Multinationals in the food and beverage sector are among the leaders in understanding the risks and opportunities presented by water, thereby driving demand for water tech. In particular the global beverage companies are acutely aware of the value of water to their businesses and are actively managing their risk. And risk management in some cases includes deploying innovative water technologies.

Corporations with large water footprints, such as beverage companies, are commercial buyers for advanced and breakthrough technologies. Many companies are focused on moving beyond just water efficiency and into "replenish" programs to drive the recycling, reuse and offsetting the amount of water they use for their operations, supply chain and in product use (the entire value chain).

The Coca-Cola Company (TCCC), for example, has publicly committed to a vision of water replenishment by 2020.[32] Companies developing leading water stewardship strategies are driving demand in technology innovation. They have the benefit of being able to overcome issues such as the current low water pricing (in some cases they use a "shadow price" for water or they can make investments as part of a broader corporate social responsibility strategy). More on shadow pricing later.

A business like TCCC finds itself swarmed with water tech vendors, but offers the promise of a motivated buyer of water tech. For a seller of water tech, that means an internal team at TCCC dedicated to water management, as well as to sustainability. That internal team has already educated itself on its needs and the water tech market's offerings. Thus, approaching a leading edge customer like TCCC can accelerate the purchasing cycle by cutting the time required to educate a customer on the benefits of water tech, and the time required to convince the customer to act because an internal mandate for water management already exists.

The recent announcement by TCCC that they have "teamed up" with Dean Kamen on deploying water tech to bring water to with world's poor is a good example of how private sector companies are driving adoption of water tech.[33] According to Reuters, TCCC plans to deliver and operate water purification systems in rural parts of the developing world, working with Dean Kamen, the inventor of the Segway transportation device, on a project that will also help further TCCC's sustainability targets.

Dean Kamen's new invention is called the Slingshot and has the goal to bring clean water to areas where it is limited. The Slingshot uses a vapor compression distillation system that runs on very low levels of electricity. Through boiling and evaporation, the system can clean and purify anything from ocean water to raw sewage, Kamen and TCCC said. One Slingshot unit can purify up to 300,000 liters of water a year, or enough daily drinking water for about 300 people, Kamen said. Kamen planned to deliver 30 Slingshot machines to TCCC by the end of 2012 and, in 2013, TCCC will place machines in rural areas of South Africa, Mexico and Paraguay, in places like schools, health clinics and community centers. By 2014, the distribution should widen to include thousands of units, and later to extend to India, the Middle East and Asia.[34]

Most businesses, however, lack an internal mandate for water management and water stewardship. Only those with either an advanced sustainability mandate, or those that use massive volumes of water in producing goods, or those that require expensive treatment of water inflows and outflows, have spent significant dollars on water tech to date.

Even among sophisticated multinationals, water tech seems early in its adoption. One Fortune 100 company that has a strong track record of leading in sustainability areas recently asked Banyan Water to assist in making the case to the company's management across different business units that the economic benefits and risks associated with water merited prioritizing the creation of an internal water mandate. As one might expect, without that mandate, the company would be ill-positioned to spend significantly on water tech. Among those that report water sustainability initiatives, companies vary significantly in their actual spending on water tech. Moreover, many companies, even those with sustainability mandates, embark on one-off projects on water management that make good publicity case studies, but fall outside of a system-wide water mandate. Thus, water tech purchasing stagnates within those customers.

One of the positive trends supporting an accelerated water tech adoption curve is the number of companies who feel they have picked the "low-hanging fruit" to reduce energy costs and risks. Most companies prioritized energy before water as for most companies, though not all, energy spend exceeded water spend. This is now changing.

The next place to turn for utility and sustainability management is water. In its State of Sustainable Business Poll in 2012, Business for Social Responsibility surveyed over 500 business leaders from over 300 companies. Those company respondents ranked water to have achieved little progress among sustainability issues over the past two decades, but projected it to be the single area of greatest progress in the coming two decades.[35]

Another trend driving increased corporate purchasing of water tech is to build the value of water into capital evaluation strategies. We discussed the externalities related to water, and the gap between water's price and its value. I (Sarni) have worked with a number of Fortune 500 companies to understand water throughout their entire supply chains. Moreover, those companies have begun to exert influence over their suppliers to take water into better consideration. As food companies reward sustainable water management practices by their agricultural suppliers, and retailers evaluate the water footprints of the goods they sell and exact rewards and punishments on suppliers for their water impact, we will witness a revolution in water management driven by corporate purchasing.

An example of how a private sector company overcomes the low price of water to make long-term investment decisions in water tech is Nestlé. Nestlé Chairman Peter Brabeck-Letmathe has repeatedly argued that more value must be put on water use through market pricing mechanisms that could drive adoption of greater efficiency measures. He has outlined three options that could help accelerate resource savings across business and industry, but cautioned that the notion of pricing was "a very delicate issue" in terms of how water is perceived. Acknowledging that pricing regulation could be one way forward for water management, he also pointed to two other possible solutions – cost curving and shadow pricing. All three were "not necessarily conflictive" he added.[36]

According to Brabeck-Letmathe:

> for every water basin of the world we are establishing a cost curve – what it would cost, what we would need to implement, in order to balance out the demand and supply side.... The third possibility is what we are doing at Nestlé. Here we have established a shadow price [for water] ... in order to see whether economically we can invest in water efficiency.[37]

The shadow price is set from one franc per cubic meter of water up to five francs per cubic meter depending on whether it is sourced from a water poor region or not. "Once you have that shadow price you then can justify the investments which are necessary to increase your water efficiency. If the water has zero value there is no economic justification for any investment," Brabeck-Letmathe explained.[38]

I (Sarni) have seen this with increasing frequency. Pricing water according to business value drives investment in water tech which is, in turn, better aligned with sustainability and water stewardship strategies. Recently I was in a meeting where a manufacturing leader commented that water risk should be factored into their process for evaluating capital investments in water tech. He had moved beyond viewing the current price of water as the key factor in evaluating capital investments. He was now thinking water risk and the real value of water to the long-term operation of facilities.

Water tech innovation is being driven by the needs of the public *and* private sectors. Historically, the industry has been driven by large public sector infrastructure projects. Many private sector organizations now recognize that water is critical for their businesses and proactively seek solutions.

Although those private sector buyers may stand at early stages of the buying process, they have the power to rapidly commercialize water tech as they move toward mass purchasing. We are both seeing this change in the marketplace at the corporate level and in buying decisions at the facility level.

Notes

1   www.globalwaterintel.com.
2   www.globalwaterintel.com/archive/12/10/analysis/opportunities-malaise.htm.
3   Interview with Tom Pokorsky for this book, 16 November 2011. Subsequent quotes from Pokorsky are all from the same interview.
4   www.epa.gov/lawsregs/laws/cwa.html.
5   Ibid.
6   https://docs.google.com/a/banyanwater.com/viewer?a=v&q=cache:1srlJcs3je8J: www.gewater.com/pdf/pr/2006_03_14_zenon.pdf+&hl=en&pid=bl&srcid=ADGEE Sj8nWA44mIGETlNXJpKHmrFtnw1qWHDXjgJjBT5OZMr36sLZOvIX3jmGmNlo lrZ2Jw72GNElEkbrDi-3mP9q5loSpIiOlZt6ZomRYe4vRzKbbgZ_DcwSm KlHZASRVNXfBStUeyV&sig=AHIEtbTDkL1xwy5c8opCWhXL78e0F2e-3A
7   www.helwan.edu.eg/TTO
8   Joshua B. Powers (2004) "R&D Funding Sources and University Technology Transfer: What Is Stimulating Universities to Be More Entrepreneurial?" *Research in Higher Education*, vol. 45, no. 1, February.
9   Adapted from http://entrepreneurshipucdavis.edu.
10   www.wipo.int/export/sites/www/freepublications/en/intproperty/928/wipo_ pub_928.pdf
11   Ibid.
12   http://niwr.net.
13   www.innovationcenteroftherockies.com.
14   www.innoviscop.com/en/definitions/innovation-clusters.
15   www.co-waterinnovation.com.

16   Ibid.
17   http://waterinnovation.net.
18   Ibid.
19   Ibid.
20   www.watercampws.uiuc.edu.
21   www.cleantechopen.org.
22   Interview with Rebeca Hwang for this book.
23   http://usa.chinadaily.com.cn/china/2012-04-20/content_15103982.htm.
24   http://seekingalpha.com/article/146151-china-s-water-crisis-is-an-investment-opportunity.
25   http://mobileet.timesofindia.com/mobile.aspx?article=yes&pageid=11&sectid=edid=&edlabel=ETD&mydateHid=02-07-2010&pubname=Economic+Times+-+Delhi&edname=&articleid=Ar01106&publabel=ET
26   www.csiro.au/en/Organisation-Structure/Flagships/Water-for-a-Healthy-Country-Flagship/India-Australia-Water-Partnership.aspx.
27   www.timesofisrael.com/i.srael-and-china-sign-billion-shekel-deal   and www.reuters.com/article/2012/02/29/israel-china-water-idUSL5E8DT4WW 20120229.
28   www.ceres.org/issues/water.
29   2030 Water Resources Group (2009) *Charting Our Water Future: Economic Frameworks to Inform Decision-Making*, www.2030waterresourcesgroup.com/water_full/Charting_Our_Water_Future_Final.pdf.
30   Ibid.
31   http://portal.luxresearchinc.com/research/report/8345.
32   www.cokecorporateresponsibility.co.uk/future-challenges/the-coca-cola-company-2020-vision.aspx.
33   "Coke, Segway inventor team up on clean water project," Reuters, 25 September 2012, http://uk.reuters.com/article/2012/09/25/cocacola-water-idUKL1E8KON6D20120925.
34   Ibid.
35   https://www.bsr.org/reports/BSR_GlobeScan_State_of_Sustainable_Business_Survey_2012.pdf.
36   "Nestlé makes case for water pricing to boost efficiency gains," 19 December 2012, www.edie.net/news/4/Nestle-makes-case-for-water-pricing-to-boost-efficiency-gains/23751.
37   Ibid.
38   Ibid.

Chapter 10

What does success look like?

If you are fortunate to live in a part of the world where water infrastructure is reliable, imagine, for a moment, your own daily dependence on water systems: your morning shower and first cup of coffee. A glass of water, just before you drink it, pumped through thousands of miles of piping and treated to rigorous regulatory compliance requirements. Rain and wastewater efficiently channeled away and treated before discharge into the environment. Think about the embedded water in your food, clothing and even the computer chip in your laptop.

What should the future of water look like? Allowing you the comfort of depending on these systems? Expanding discretionary comforts and basic thirst quenching and sanitation to those who need them? Reducing the impact of water systems on the environment?

Success, in twenty-first-century water management, means more than enabling the status quo for another hundred years. It means transforming water supply and use to a sustainable model in balance with the environment and ecosystems, providing for human use and providing access to sanitation to all. Moreover, success means achieving that vision while adapting to constant changes in the environment, politics, and economics.

The flourishing of water tech will carry us toward success along with changes in water pricing, public policy and "collective action" initiatives. Throughout this book, we have discussed the myriad challenges to this vision of success. Naysayers abound in every key stakeholder group. Yet, we believe the water tech train has left the station. We believe, in the very near term, we will see a dramatic increase in the adoption of water tech, in the demand for water tech, and in the organizations and people that enable next generation water tech to permeate the fabric of society.

Success will unfold over time, with capacity-building for innovation and policy, with new inventions, with accelerated purchasing of existing water tech by educated, motivated buyers. Perhaps best of all, because of the very obstacles that have blocked water tech from more rapid creation and adoption, we are still early. That means each of us can make an impact.

Let's briefly examine the agricultural sector to envision what success looks like.

What might the agricultural sector in a twenty-first-century water paradigm look like? The agricultural sector understands water. Our twenty-first-century water-smart farm has an even deeper level of readily available water information, and a set of tools and best practices to use water efficiently, to control waste runoff, and to integrate with the broader ecosystem. Our water-smart farm may receive its supply from an irrigation district. That district would monitor water supply, and base its price and availability economically to create efficient use incentives for our farm. If the irrigation district delivers water through canals, it would deploy methods of harnessing power from water flows in that canal to reduce the cost of delivering water, and the energy and carbon impact of each gallon of water our farm uses. Our water-smart farm has an advanced data management system, including data tracking and benchmarking of water requirements for each crop to supplement our farmer's experiential understanding of water-plant interaction. Our farm uses an efficient method of irrigation. Our team evaluates precision agriculture water tech, from polymers that absorb water in soil to reduce watering requirements, to a sensor system in the field to detect soil moisture, salinity, and other variables. Our advanced water tech maximizes crop per drop, and may also enhance our crop yield. Our farm monitors runoff and potential contaminants from waste. We minimize waste, and evaluate water tech to enable waste-to-value opportunities. Our water has market value, and we have access to market-based systems to maximize that value, from storage to resale. Our water-smart farm integrates not only with our water supply district, but also with our buyers' supply chain. The crops we sell are water-smart crops – drought and salt water tolerant where needed with a quantified embedded water footprint. Our buyers and, in turn, retail companies reward us for these best practices with increased demand and premium pricing. Finally, we participate in best-practice working groups to share our water management methods, and learn from others using advanced water tech.

What will a water-smart city look like?

Most cities will, in the coming years, retrofit themselves with water tech. What might an extreme urban water makeover look like? For a glimpse, let's explore the water management practices of a new city, built as a paradigm for sustainable development.

Masdar City in Abu Dhabi is a master-planned model for technology innovation in water, energy, green building and urban transportation. Water innovation is essential to the success of Masdar City, along with a focus on six key areas: smart grid and building automation, advanced cooling, lighting, sustainable materials, energy management software, and electric mobility. Brian Fan, Masdar City's head of research, technology roadmap and commercial development, is thus at the center of crafting one version of a twenty-first-century water paradigm.[1]

Masdar City, with a budget and a mission to build this paradigm, is an emerging global hub for renewable energy and clean technologies, offering a home for leading edge research and development, pilot projects, technology testing, and construction on some of the world's most sustainable buildings. Masdar City serves as an open technology platform that gives companies an opportunity to develop, test and validate their technologies in a large scale, real-world environment – and in particular, with consideration to the region's climate conditions and consumption patterns.

Out of necessity, water innovation is embedded into the design and operation of Masdar City. Its water requirements indicate how much we may achieve with water tech. The water needs for Masdar City are less than half of a "business as usual" approach. In Phase 1 of the project, Masdar City designers are targeting consumption of 180 liters per person per day, well below the business-as-usual rate of 550 liters per person per day. The city's target will be progressively lowered, to an ultimate goal at full build-out of 40 percent below the Phase 1 target.[2]

To achieve these lower consumption figures, the city is using a range of water tech. Already, the city uses highly efficient fittings, fixtures and appliances, and smart water meters that inform consumers of their consumption and use data across the entire network of meters to identify leakage across the system. As the city pursues more ambitious consumption targets, additional strategies will be implemented, including a water tariff to promote further water efficiencies.

Outdoor landscaping often comprises 50 percent of urban water use, of which often half or more represents inefficient waste. Moreover, in most cities, landscape water has already been treated to drinking water standards. The water tech vision of the future overcomes these wastes. In Masdar City, treated wastewater is 100 percent recycled for use in landscaping. Masdar City has also has achieved a 60 percent reduction in landscape water usage of that recycled water per square meter over business as usual, through a variety of strategies, including highly efficient micro-irrigation, landscaping design that minimizes plant evapotranspiration, and low-water-use and indigenous plants and trees.[3]

Water tech urban design is in process, *now*, in Masdar City.

Which brings us to an essential part of what success looks like: *timing*. Water risk and business opportunities are here and now, not decades in the future. Although we believe the vision of a water tech future will come together in pieces, with some technologies and practices advancing faster, and some customers valuing water sooner, we believe real progress will come in months and years, not decades.

It is easy to become complacent about water risk and business opportunities. However, water scarcity is disrupting businesses in the US and elsewhere in the world and companies are developing technologies to transform how we manage water.

Peter Brabeck-Letmathe, Chairman of Nestlé, noted in a 2013 interview at the World Economic Forum in Davos that:

> five years ago I started to talk about the water issue and we were ten people in the room, today water is being recognized as the number two societal risk in the world. That's quite a good step and I am gladdened by the amount of companies and colleagues of mine who are getting involved.[4]

Indeed, as previously discussed, the results of the 2012 CDP Water Program indicate that the new age of water tech may be upon us. Most respondents to the questionnaire recognized they *were currently exposed* to water related risks and were *already being impacted*. Equally important is that the majority of companies have identified business opportunities in addressing water related risks, from cost savings to the development of new products and services.

Three other reports released in 2012 also identified water risks as a current issue. A report by the World Economic Forum identified water risk as *high likelihood and high impact* (*Global Risks Landscape 2012* report).[5] In addition, reports by the Economist Intelligence Unit[6] and Ceres[7] identified how water scarcity represents a *current risk* to water utilities.

Certainly, reports such as these don't always translate into action. If the immediate economic impact of action is obscured or lacking, then actors are prone to deferring investment. We have witnessed similar delays in other climate and sustainability areas. When I (Sarni) was having conversations with businesses on climate change risks in the late 1990s it was viewed as a risk that could be addressed at a later date (if climate change was viewed as a real risk at all). There was little sense of urgency to manage climate risk or develop technologies to address this risk. Over a decade later, much has changed. Look at how many businesses and governments have a climate change/energy strategy and how the clean tech sector has emerged in response.

Most businesses can ignore water risks and opportunities for the moment. Those that act now, however, can develop or buy new products and services to become resilient to water risks and business disruption, and to capitalize on water opportunities. We see companies, homeowners, farmers, utilities and governments beginning to act, indicating approaching widespread change.

Which leads us to what it will look like as stakeholders *take action*.

Building the twenty-first-century water industry *will take coordinated action* in several key areas. An ad hoc approach to innovation in the water industry has not worked well and will clearly not work going forward.

What will it take to create the twenty-first-century water industry?

While we have covered these issues previously it is worthwhile to recap what will be required to move us off the business-as-usual trajectory. To summarize, there will need to be action in the following key areas.

Education to create awareness and action on the importance and value of water

Where do we begin in an effort to spark widespread adoption of a new water paradigm? Educating key stakeholders to make them aware of the water problems and risks they face, but more importantly the water opportunities available, is an important early step to inspiring action.

Most people do not know how much they pay for water or sewer, or how their bill works. Nor do most companies. Most governments consider water and sewer a necessary public service more than an opportunity to build a pillar of economic growth. Most entrepreneurs, innovators, students and academics think of water as a near-free commodity provided by their town, rather than a business, no less a massive industry open to disruption.

In countries with reliable access to water, part of the fun of talking to people about water is watching their eyes open as they become aware of our current water situation, and watching their eyes open a little more as they learn of the opportunity to make a change.

Stakeholder education on water tech can and will occur in multiple forums, from primary schools to higher education, to utility and city programs, to trade shows and government conferences. Many education opportunities already exist. So what will be different, beyond more classes and more opportunities? What will actually change perception and behavior?

One part of that answer goes beyond education, to the other pillars of the twenty-first-century water paradigm we lay out below. As noted below, regulation and pricing to the true value of water can force stakeholders to get smart on water, which will then lead to action. Without changing economic forces like regulation and price, we can still inspire stakeholders to act with information. Transparency into the current system makes a big difference. For example, as noted in the 2012 Xylem study of public attitudes on water pricing (discussed in Chapter 1), US consumers are willing to pay for investment in water infrastructure but the linkage between their personal use and value is not always apparent.[8] Another key lies in highlighting the value of teaching and learning about next generation water management, which in turn depends on turning stakeholders on to the water opportunity.

In addition, the delivery of the message can be as important as the content. Cities can educate stakeholders on the value of reused water, for example, only to fail in inspiring citizens and businesses to accept water reuse as a practice. Alternatively, with smart marketing and branding, cities can undertake the same educational effort but succeed in inspiring a new water paradigm. In the US, several attempts to overcome the stigma of reused water have struggled with "toilet to tap" branding, including programs in the Orange County Water District in California (groundwater replenishment system) and in San Diego, California. Each of these programs, despite significant stakeholder education efforts, met public pushback from the "ick factor." Conversely, Singapore has coupled sustainable water use education with smart branding to rebrand reused water as "New Water." Singapore now bottles and sells "New Water," and has found many industrial customers requesting reused water because they have found better consistency of that water than regular tap water. These examples highlight the need for education, especially education that demonstrates the opportunity of better water management, to inspire action.

Building a water ecosystem through collective action

The twenty-first-century water industry will be built through a "water ecosystem" to promote innovative thinking and nurture companies to transform how we manage water. The development of a more robust water ecosystem will require public and private sector cooperation through collective action.

The water ecosystem is, and will continue to be, a driver in moving us to a new way of thinking about water and the development of new "disruptive" water technologies. The Blue Economy Initiative in Ontario, Canada is an excellent example of a successful water ecosystem (as previously discussed, there are several others, including Singapore and Israel). The groundwork for the "Blue Water Economy Initiative" began with the 2010 *The Water Opportunity for Ontario*[9] report and the *Blue Economy: Risks and Opportunities in Addressing the Global Water Crisis* report.[10] These reports suggested that Ontario, Canada act on an opportunity to become a world leader in the massive global water market. The former report was the effort of about 100 companies in Ontario and included a wide range of stakeholders – academics, industry, consultants and government representatives.

The recommendations from this report highlight the key elements of creating a successful water ecosystem (the recommendations outlined below are specific to Ontario but several elements can be applied anywhere):

- *Establish a bold vision* – this is all about strong leadership and vision. The report takes the position that Ontario will need to send a "clear unifying message that water is a high priority, now and for the future." The goal, by 2015, called for Ontario to be "recognized as a global leader and centre of expertise for providing safe, clean, affordable and sustainable water solutions."
- *Create an Ontario Sustainable Water Opportunity Act* – align public policy with private sector entrepreneurship. According to the report, Ontario has become a global leader in energy conservation and renewable energy in part due to the introduction of the Ontario "Green Energy Act, 2009." The creation of the water opportunity act is believed would: encourage sustainable water behavior; adopt transparent costing and accounting of water use; support water technology demonstration and early adoption; and attract early stage, innovative water technology companies to Ontario. The policy would tie public funding of utilities to demonstrated adoption of new water tech.
- *Increase alignment and collaboration* – although the value of aligning stakeholders and promoting collaboration may be straightforward, fragmentation and misaligned incentives inhibit cooperation on water management globally. The report recognizes the market's fragmentation, the need to identify and address market barriers, and the need to engage all stakeholders to create opportunities.

- *Brand Ontario as a leader in sustainable water* – brand matters. The report recommended that Ontario develop and market its brand. In building Ontario's brand, a clear articulation of Ontario's "unique water story, both within and outside Ontario, would increase the perceived value of Ontario's water industry, products and services."

The pillars of the Ontario report can be the pillars of a broader, global water ecosystem. Painting a big vision is a necessary step toward gathering and motivating the key stakeholders for collective action. In Ontario's case, the authors recognized the major economic potential of water tech, and aimed high at demonstrating how important it could be to the regional economy. The authors also recognized the existing market impediments to a flourishing water economy. In the spirit of the report's big vision, however, the authors also put forth a firm belief that the water market would grow into an increasingly important economic, political, and social focal point for the region and the world. Educating key stakeholders to the current opportunity, and painting a big vision for how that opportunity will grow, are key components of inspiring water ecosystems for collective action.

As discussed in Part I, Ontario's vision is one of a number of efforts to build water ecosystem hubs and organizations. Other organizations shaping the twenty-first-century water paradigm (such as Imagine H2O and Kinrot Ventures) united various stakeholder groups to build water ecosystems. While the few existing initiatives to build water ecosystems vary, they share the goal of connecting a diversity of stakeholders to identify and nurture water tech innovation. Although we believe water tech ecosystems will interconnect, enabling a broader global platform for next generation water management, local and regional efforts will remain crucial. Water, for now, is a hyper local issue. Building local efforts that then interconnect where necessary can help, and multiple efforts can create both opportunities for partnership and for healthy competition for leadership, advancing water tech at a quicker pace.

Smart public policy

The private sector can make great strides in implementing water stewardship strategies and developing innovative water technologies. However, real long-term progress can only go so far without the support of smart public water policies. Water is, after all, a heavily regulated commodity.

As previously discussed, smart public policy starts with smart water pricing. As we have discussed, the price of water does not reflect the full cost and certainly does not reflect the full value of water. Almost everyone in the water industry echoes the disconnect between current pricing and the need for smart water pricing. As long as water prices are artificially low, there will continue to be a built-in disincentive for investment in water tech. To quote a colleague, the "tyranny of the simple payback" is a real barrier for investment in water projects. While the private sector can reduce the payback period for investment in water projects and institute a "shadow price (the setting of an internal price for water independent of local water prices to drive capital investment)" on water, the public sector is more challenged.

Smart water pricing means more than increasing water and sewer prices where they fall too far below true all-in cost to create good behavior. It means smart methodology of pricing, ensuring affordable service and supply for basic needs, and economic pricing of marginal system use. It means smart measurement of use to enable better pricing, smart billing to enable better collection, and much more. Creating a better twenty-first-century water management system requires placing appropriate value on water, and choosing economic methods to provide for basic needs and to provide appropriate incentives for smart water behavior beyond those basic needs.

In our twenty-first-century water paradigm, smart policy applies to all water, including wastewater and reused water, and we believe we will see increases in price and value across the board. In part, we believe those changes will result from policy which encourages matching water quality to water use, and encourages reuse. Currently, many developed countries use water treated to drinking standards for all uses, and use that water only once, whereas peer developed countries reuse the vast majority of their water. We expect smart policy, incorporating price, regulation, building codes and more to encourage "catch up" from regions which employ a "use it once" or a "use it or lose it" model. We will move toward a policy framework in which all water is valued.

Moreover, smart water policy transcends smart water pricing. We have discussed the importance of policy promoting water tech supply and demand to enable a smart water future. Smart water policy will also support and link the development of water ecosystems, in part by encouraging public and private collaboration. Some existing platforms for public private partnerships to tackle long-term water stewardship and smart policy include efforts by the Water Resources Group, WWF and the World Business Council for Sustainable Development (WBCSD). All of these efforts are making progress in getting public policy to move towards a new long-term approach to water management. In addition, there are numerous examples of public-private cooperation on financing water tech and water research, all of which provide a key second pillar to a vision of better water management.

Advanced water policy may contemplate water in its interaction with other basic pillars of society, recognizing and supporting the link between water, food, and energy along with the needs of ecosystems, social uses and cultural norms. Increasingly, the private sector has engaged the public sector on aligning energy, water and food policies. As we have seen, water scarcity impacts energy and food production. Smart water policy, as it relates to energy and food production, may address water efficiency programs, promoting low water footprint energy production in water stressed watersheds, innovative water reuse approaches for shale gas production, and growing crops appropriate for the region and watershed, among other linkages.

Smart capital

Building a new water paradigm requires capital. We believe we will see increased investment in water opportunities from a growing investor base who will rapidly climb the industry learning curve. We believe those investors will benefit from a growing wealth of qualified investment opportunities managed and dreamed up by teams of operators and entrepreneurs, some of whom will come from the industry, and many who will be drawn to the industry as we educate more stakeholders to the water opportunity.

Supportive regulation and policy will facilitate new capital for smart water initiatives, as will increasing water and sewer prices, and the incorporation of water risks and externalities into customer perception of water value.

We expect the financial community to play a part in accelerating water tech production and adoption by performing diligence on water risks and opportunities, and pressing seekers of capital to address water needs. Debt investors will increasingly investigate the required water spend of bond issuers, forcing those cities and utilities to spend on water tech. Public equity investors will expect companies to report on water use and act upon opportunities to save money and reduce risk by adopting water tech. Private market investors will offer to finance improvements in critical water systems along with more economically-driven water and sewer pricing methods, and will seek to invest behind high return water tech improvements in those systems. Cities will tie their own financing of water infrastructure to advanced water tech.

A water-smart future requires a virtuous cycle of capital and entrepreneurship. We believe water-smart capital will rise in dollars, sophistication, and impact.

Leadership development

Students and professionals interested in entering the water market frequently ask if we believe water will be "the next big thing." Our answer, often, is "yes, if you make it so." To move beyond the old paradigm of water management, we need to attract and develop a new generation of water leaders in all relevant stakeholder groups, and those leaders must be comprised of new entrants as well as new leaders emerging from existing ranks.

This will mean moving public sector leaders to fully understand water risk, fair water pricing and technology innovation, and to encourage technology innovation through acceptable risk taking. The current mindset of "just do it the same as before" will not serve us well going forward. Public sector leaders will need to reject accepting non-revenue water, highly subsidized water pricing and a supply side mentality (the focus on increasing supply as the default solution to water scarcity). That will require inspiring new leaders within public sector ranks, and attracting new leaders to advance water initiatives from outside existing ranks.

The private sector will also need to develop leaders well versed in water issues and solutions. We have seen considerable progress as legal, MBA and MS programs have come to include sustainability courses. However, water steward-ship now needs to be embedded into these programs just as climate change and renewable energy issues have been to date. For those already in leadership posi-tions in the private sector, internalizing externalities such as water needs to be the norm and not the exception.

Fortunately, we are seeing this shift now as young leaders are emerging in water tech and water stewardship strategy. We have found no shortage of students and professionals interested in working in water and on supporting water stewardship strategy projects.

Each of these pillars links to and will support the others. Together, education of key stakeholders, ecosystems linking those stakeholders, supportive policy, smart capital, and a new generation of leadership will build a vision of success.

Notes

1 http://masdarcity.ae/en.
2 Ibid.
3 Ibid.
4 www.guardian.co.uk/sustainable-business/nestle-peter-brabeck-attitude-water-change-stewardship.
5 World Economic Forum (2012) *Global Risks 2012*, http://reports.weforum.org/global-risks-2012.
6 Economist Intelligence Unit (2012) *Water for all? A Study of Water Utilities' Preparedness to Meet Supply Challenges to 2030*, www.oracle.com/us/industries/ utili-ties/utilities-water-for-all-ar-1865053.pdf.
7 Ceres (2012) *Water Ripples: Expanding Risks for US Water Providers*, Boston, MA: Ceres.
8 Xyleminc (2012) *Value of Water Index – Americans on the US Water Crisis*, www.xyleminc.com.
9 OCETA and XPV Capital Corporation (2010)*The Water Opportunity for Ontario*, March.
10 David Henderson of XPV Capital Corporation and Nicholas Parker of the Cleantech Group (2012) *Blue Economy: Risks and Opportunities in Addressing the Global Water Crisis*, www.blue-economy.ca/sites/default/files/reports/BlueEconomyInitiative%20-%20WEB.pdf.

Chapter 11

The Global Water Forum 2024

Having been to the 2012 Global World Water Forum, I (Sarni) can picture what the 10th World Water Forum in 2024 might look like. From 2012 to 2024 there has been slow but measurable progress in addressing water scarcity and water quality issues.

In 2024 the conversations with CEOs, public sector leaders and NGOs are more positive than they were in 2012. By adopting water stewardship strategies, actively engaging in collective action programs and deploying innovative technologies, we have benefitted from increased access to clean water and sanitation, economic growth and the adoption of "fair" water pricing to drive conservation and innovation.

I believe 2012 will be viewed as a turning point in addressing water risk and driving technology innovation. The CDP Water Program reports, the World Economic Forum report on Global Risks Landscape (identifying "water supply crisis" as having a high probability of likelihood and high impact) and the drought in the US and elsewhere brought us to the tipping point.

One can imagine a conversation with the CEO of a global food company during the 2024 forum reflecting on the progress the private and public sectors have made over the years. Her company's supply chain is efficiently using water resources through adoption of technology and sustainable agriculture practices. Water scarcity and water quality are not limiting her expansion into many parts of the world giving her the ability to feed an ever-increasing population. She knows the water footprint of her products and communicates the performance of her company's water stewardship program through frameworks such as CDP (now expanded to cover all natural resource issues – energy, water, ecosystems). Her company is an active partner with the public sector, NGOs, and other stakeholders in deploying water technology to improve water efficiency, reuse, and recycling.

I can also picture several government representatives being proud of their accomplishments: innovative water pricing tied to the value of water; restructuring of their water and energy utilities to minimize water use (consumptive and non-consumptive) through conservation; smart agricultural policy to promote growing low water requirement crops in arid and semi-arid climates; deployment of low water energy production and the creation of water technology hubs to build a thriving technology export industry and a robust educational system creating generations of water engineers, scientists, public policy leaders and entrepreneurs.

And finally, the NGOs have made significant contributions in addressing water issues. They have led in the development of water footprinting methodologies, water risk mapping tools, collective action platforms such as the Water Action Hub, and supporting public policy in ecosystem conservation and transboundary water stewardship. These NGOs have truly become partners with the private and public sectors in deploying innovative water policies and technologies.

I am hopeful that 2024 will be viewed as a great meeting and that we can be proud that businesses, the public sector and NGOs heeded the call to action over 10 years ago. The "business as usual" scenarios did not come to pass. Instead, we have come to value water and have new technologies and partnerships successfully addressing the public sector, private sector and ecosystem needs for water.

Looking back a decade, both of us (Sarni and Pechet) will be struck by how present the opportunity to make a difference in building our water future was, and be grateful that we were among those to learn of it and act upon it.

Index

9781032926681